LEFT ELSEWHERE

FINDING THE FUTURE IN
RADICAL RURAL AMERICA

Editors-in-Chief Deborah Chasman & Joshua Cohen

Executive Editor Chloe Fox

Managing Editor Adam McGee

Senior Editor Matt Lord

Engagement Editor Rosie Gillies

Editorial Assistant Spencer Quong

Publisher Louisa Daniels Kearney

Marketing and Development Manager Dan Manchon

Finance Manager Anthony DeMusis III

Distributor The MIT Press, Cambridge, Massachusetts, and London, England

Printer Sheridan PA

Board of Advisors Derek Schrier (chairman), Archon Fung, Deborah Fung, Alexandra Robert Gordon, Richard M. Locke, Jeff Mayersohn, Jennifer Moses, Scott Nielsen, Robert Pollin, Rob Reich, Hiram Samel, Kim Malone Scott

Interior Graphic Design Zak Jensen

Cover Design Alex Camlin

Left Elsewhere is *Boston Review* Forum 9 (44.1)

To become a member, visit:
bostonreview.net/membership/

For questions about donations and major gifts,
contact: Dan Manchon, dan@bostonreview.net

For questions about memberships, call 877-406-2443
or email Customer_Service@bostonreview.info.

Boston Review
PO Box 425786, Cambridge, MA 02142
617-324-1360

ISSN: 0734-2306 / ISBN: 978-1-946511-40-9

Editors' Note

Deborah Chasman & Joshua Cohen

THE CARTOGRAPHY OF U.S. POLITICS has hardened into cliché: islands of urban blue in a vast sea of rural red. In the recent midterm elections, however, rural working-class whites shifted their votes toward Democrats by seven points. (Rural people of color already largely vote Democratic.) To appreciate the magnitude of the shift, consider that Democrats picked up only two points among blue-collar suburbanites. Still, winning in rural districts will require the Democrats to make a fundamental strategic choice: Do you mobilize voters with a boldly progressive platform, or do you moderate the message in the hope of swinging some independents and Republicans?

From her vantage point in the Shenandoah Valley, historian Elizabeth Catte, our lead essayist, sees "a failure of imagination" in the way Democrats are approaching this strategic issue. Catte argues that a deep strain of rural radicalism is already working—with some notable successes—on environmental issues and the fight for a living wage. Moreover, she thinks mainstream, urban progressives have been on a fool's errand to try to understand the mysteries of red-state "Trump

Country." "It would be far better," Catte says, "for progressives to save their bubble-popping for moments such as this, when an opportunity emerges to better understand those closer on the political spectrum in those same spaces. The limbo we are trapped in compels white progressives to see themselves in a Trump voter rather than in a rural socialist or communist."

Left Elsewhere grabs this opportunity by putting rural progressives in conversation with their urban cousins. The responses to Catte are varied and passionate. Taken together, they represent a compelling engagement with this important political challenge.

Other essays in the issue provide context for this engagement. In rural Ohio, we see an educator conflicted about being recruited by his district to carry a concealed gun at school. In Louisiana and Mississippi, Robin McDowell and Makani Themba share stories of contemporary organizing informed by the long history of black resistance in the South. In Tennessee and Kentucky, Lesly-Marie Buer shows how radical tools first developed by marginalized groups are proving most successful in combatting the opioid crisis. And in North Carolina, William Barber explains why "the South holds the key to transformation in this country."

Left Elsewhere was prompted by an immediate political challenge. But it opens up a larger set of issues about how to rethink U.S. politics by linking rural and urban values and experiences. It starts a conversation that will, we hope, continue long past 2020.

—December 21, 2018

Chasman & Cohen

Left Elsewhere

Elizabeth Catte

WHEN MY GRANDFATHER was a child, his stepfather would bring him along as he sold moonshine to poor working men in southwestern Virginia coal country. The men adored my grandfather, who was not yet even school age, for his talent mocking Democrats. He told me this story on a few occasions to explain, I think, the inevitability of his later affiliation with the Republican Party. He was a Republican in much the same way that I am a Democrat—voting a ticket with little enthusiasm every few years and sometimes not at all.

When I consider that story now, I find myself thinking less about my grandfather and more about the men who laughed at his jokes. What were their politics? Not all were the predecessors of today's Republicans, as we might imagine them to be. In Appalachia, so-called "mountain Republicans" comprised an old vanguard of anti-secessionists who thought of themselves as particularly enlightened—heirs, they imagined, to the legacy of Abraham Lincoln. My grandfather belonged (or at least aspired to belong) to that tradition. His audience might have consisted of Democrats, who enjoyed hearing their abuses repeated in the mouth of a child. But it is more likely that they would describe themselves as without politics, just laughing at the powerful and self-important. For a long time, it did not occur to me there were other possibilities.

My wider view of politics in Virginia's coal country changed when I discovered that the publisher of my grandfather's local community paper, *Crawford's Weekly*, was a communist. And not just a communist in print, but a shot-while-inciting-class-war, sabotage-the-New-Deal-from-within, run-for-local-political-office-on-a-platform-of-a-producer's-republic communist. His name was Bruce Crawford, and when my partner, also from southwestern Virginia, discovered his writings, we read them aloud to each other as though they were letters from an eccentric uncle.

Our favorite piece of his writing comes from the pages of the *New Masses*, a U.S. Marxist magazine that flourished between the world wars, where he announced in 1935 that he had killed his own paper because it interfered with his politics. "It was too radical for its bourgeois customers," Crawford wrote from Norton, Virginia, "and not radical enough for me. Like capitalism, it was full of contradictions. Hence it could not go on." The essay, "Why I Quit Liberalism," is an exceptional piece of early #quitlit, with the same indulgent qualities. "If I get shot in the leg again, or go to jail, there won't be that damned feeling of apology to the respectable," he wrote. "With the more tangible roots to bourgeois life severed, I hope to know a new and meaningful freedom, whatever the hardships."

Rural spaces are often thought of as places absent of things, from people of color to modern amenities to radical politics. The truth, as usual, is more complicated. The parents and grandparents of my childhood friends were union organizers; when my grandfather moved to East Tennessee, he went from a world of communist coal miners to the backyard of one of the most important incubators of the civil rights movement, the Highlander Research and Education Center. I now organize with people whose families have fought against economic exploitation for generations. From my vantage point in West Virginia and southwestern Virginia, what is old is new again: the revival of a labor movement, the fight against extractive capitalism, the struggle against corporate money in politics, and the continuation of women's grassroots leadership.

The question of whether mainstream liberal opinion is shifting further left has been hotly debated in the national press after Alexandria Ocasio-Cortez won the primary for New York's fourteenth congressional district with grassroots momentum and a socialist-friendly platform. Both conservative and liberal commentators predicted disaster, framing the twenty-eight-year-old rising political star as a gift to Donald Trump. Former Democratic congressman–turned–political pundit Steve Israel warned, "A message that resonates in downtown Brooklyn, New York, could backfire in Brooklyn, Iowa." Nancy Pelosi waved off the win as a district-specific what-happens-in-the-Bronx-stays-in-the-Bronx phenomenon. A few months later, Ocasio-Cortez became the youngest woman elected to Congress.

Political veterans such as Pelosi and Israel think that the cornerstones of the emerging left platform—housing as a human right, criminal justice reform, Medicare for all, tuition-free public colleges and trade schools, a federal jobs guarantee, abolition of U.S. Immigration and Customs Enforcement and for-profit prisons, campaign finance reform, and a Green New Deal—might perform well in urban centers but not so much elsewhere. Appalachia has become symbolic of the forces that gave us Trump. After all, his pandering to white racial anxiety did find purchase here. His fantasies to make America great again center on our dying coal industry. And the region's conservative voters, who have been profiled endlessly, have been a reliable stand-in for all Trump voters, absorbing the outrage of progressive readers. But what Pelosi and Israel see as common sense and pragmatism can also be interpreted as tired oversimplifications and a failure of imagination.

We remain attached, after all, to narratives that have worked very hard to simplify and neatly divide the state of the union: blue cities, red rural areas, a few swing suburbs. "In a period of political tumult, we grasp for quick certainties," sociologist Arlie Russell Hochschild writes in *Strangers in Their Own Land* (2016). Indeed, the biggest gift that the left has given the right since 2016 is not a few avowed socialists but the myth that Trump voters are inscrutable and monolithic. "I love Cleveland, but I've always considered it separate from Ohio," resident Julie Goulis told the *Guardian* just before the 2017 inauguration. "Some of the soul-searching I've been doing after the election has been about how I can understand people outside of my bubble."

It would be far better for progressives to save their bubble-popping for moments such as this, when an opportunity emerges to better understand those closer on the political spectrum in those same spaces. The limbo we are trapped in compels white progressives to see themselves in a Trump voter rather than in a rural socialist or communist—or even a rural person of color, who faces many of the same struggles as the Trump voter, perhaps even more pronounced, and chooses a different way forward.

Let us get free of that, once and for all. Appalachia should not be seen as a liability to the left, a place that time and progress forgot. The past itself is not a negative asset. The hierarchies and systems of power here feel old because they are, but this legacy also means there are many who are well practiced in the art of survival and resistance.

Our present can be reckoned with, and a different future emerge, but the way forward for the left, in my world, is through the past.

WHEN OCASIO-CORTEZ ASKS if voters are prepared to choose people over money, I hear echoes of a much older question that still resonates in Appalachia: Which side are you on? In 1931, when Black Mountain Coal Company cut miners' wages in Harlan, Kentucky, a long strike ensued. Harlan's infamously corrupt sheriff, J. H. Blair, terrorized union families; law enforcement, including the National Guard, intervened on behalf of the interests of coal operators to force miners—through threats, coercion, and violence—to return to work. When the sheriff and deputized coal company operatives ransacked activist Florence Reece's home in search of her husband, who helped organize the strike, Reece penned what would become one of history's most recognizable labor anthems, "Which Side Are You On?" The song galvanized workers and inspired bystanders to surrender the illusion that one could be impartial in the face of so much oppression. "Us poor folks haven't got a chance unless we organize," she sang. "They say in Harlan County there are no neutrals there."

Reece set her humanity against the plundering coal bosses who controlled every aspect of her family's life. Striking at the legitimacy of the ruling class, her question became immortal, useful not only to workers but, later, to civil rights activists. Would a candidate with Ocasio-Cortez's platform fail in Appalachia? Perhaps. But people would find themselves animated to hear old questions in a new context, attached to new possibilities. In fact, some already have.

In late 2016, for example, a young man named Nic Smith, another product of southwestern Virginia, made headlines for his participation in a #Fightfor15 demonstration. Outside a Richmond McDonald's, Smith, a Waffle House employee, connected the plight of fast-food workers with the past struggles of coal miners in his family. He also pushed back against the Trumpian reactionary politics that elevates white working-class racial anxiety over class solidarity. "Ain't no damn immigrant stole a coal job," Smith said. "I'll tell you that right now. And really, even if they did, would you really be blaming the immigrants or the people that hired them? The only reason they would hire an immigrant over an American citizen is if it benefits their wallets."

Instead of rigging a dying industry, Smith explains in a *Washington Post* op-ed, it would be far better to unionize low-wage workers and raise the minimum wage. He joined #Fightfor15, he wrote, because his family "has always understood that we can't wait for a savior at the ballot box to shepherd in the change we so desperately need."

A self-described "damn white trash hillbilly from the holler," Smith is an exotic figure to the many media outlets that covered him. VICE complimented him for not fitting "the image of the typical millennial activist"—a former factory worker, you see, who "isn't the kind of Democratic Socialist who spouts off at Brooklyn parties about the 'means of production.'" Smith's approach is fairly typical, however, if you are looking from Appalachia rather than New York. Here, activists such as Smith often connect to the struggles of their parents and grandparents as they engage in activism.

For decades, this distinct motivation has been at the heart of much of the success that the left has seen elsewhere. Helen Lewis, a beloved Appalachian educator who became an activist in the 1940s, taught poor people economic history to prime them for organizing. She began by asking them what their grandparents did for a living, then what their parents did. This strategy inspired generational thinking, and, according to historians Jessica Wilkerson and David P. Cline, Lewis's bottom-up

Appalachian Studies "influenced a cadre of activists, including grassroots leaders and white civil rights activists who migrated to the mountains to build alliances with rural whites." Learning about the region's history, whether through one's family or formal study, is often a crucial step in helping people understand that their struggles are a new battle in an old war.

Brandon Wolford, for example, a teacher from Mingo County, West Virginia, grew up watching news footage of 1980s miners' strikes that his father participated in. The price of coal had declined, and companies such as the Pittston Coal Company tried to recoup their losses by slashing workers' wages and benefits. In the case of Pittston, the United Mine Workers of America eventually prevailed, winning back many protections and securing others, although coal's economic slump made it difficult for organized labor to rebound fully. The memories of those and other labor victories, however, stayed with families. "Knowing the role my father and both grandfathers had played in these events sparked a special interest," Wolford wrote in *55 Strong: Inside the West Virginia Teachers' Strike* (2018). "I wanted to be involved in a movement like that someday."

In February of 2018, Wolford joined more than 20,000 West Virginia teachers and public school employees who went on strike to force state leaders to reckon with inflated insurance premiums, low pay, and a widespread teacher shortage. The strike closed schools in all 55 counties for 9 days and endured even after union support stopped. It ended with a 5 percent pay raise across the board and a pledge from state leaders to reexamine the state's public employee insurance agency.

The West Virginia teachers' strike emerged as one of the clearest visions of the new labor movement. It inspired education strikes in other states, including Kentucky and North Carolina. But understanding the strike requires knowing a century of southern West Virginia history, notably its infamous labor uprisings, from the Mine Wars of the 1920s to big coal's union-busting campaigns of the 1980s. When momentum to strike built in early 2018, teachers in West Virginia's coal country were among the first to mobilize and put action to a vote. In a Facebook group, they used coal country's labor history to portray the strike as not only urgent and just, but also natural—something that people like them had been doing for generations. In a speech at a countywide meeting, Katie Endicott, an English teacher from Mingo County, emphasized the familiarity: "If we can do this, if we can stand, then we know that our brothers and sisters in Wyoming [County] are not going to let us stand alone. We know that our brothers and sisters in Logan County will not let us stand alone. The south *will* stand. And if the south stands, the rest of the state will follow our lead."

To create solidarity in the present, to make change for the future, West Virginians needed to remember their radical past. To the extent that collective action requires a public narrative—a story that helps consolidate its moving parts and moral purpose—West Virginia's south played its part exceptionally.

THIS PAST SITS UNEASILY within Joe Manchin's vision of West Virginia. In the aftermath of the 2016 presidential election, Manchin, the state's former governor and second-term Democratic senator, became a favorite of pundits. They predicted that Democrats' only lifeline would be replicating his limp style of centrist politics. An *Atlantic* essay titled "What Joe Manchin Can Teach Democrats" touted Manchin's utility as "a sounding board for, and bridge between, his party's leadership and conservative, rural, white voters." Manchin's vote to confirm Brett Kavanaugh to the Supreme Court amid allegations of sexual assault did little to dim pundits' rose-colored view of the great moderate. Manchin was better than no Democrat at all, they reasoned—the best West Virginia can do.

Yet as writer Aaron Bady, a West Virginia native, counters, "It's because I am burdened with a handful of facts about West Virginia —and a memory that goes back more than two years—that this kind of analysis stands out as the garbage that it is." One of these facts is the 1996 gubernatorial race, in which Manchin was outflanked on the left in the Democratic primary by Charlotte Pritt. Pritt's anti-corporate platform alienated coal industry power players; Manchin, a coal millionaire, convinced a coalition of Democrats to finance and support Pritt's Republican challenger, Cecil Underwood. For many observers, Manchin's vengeance went beyond politics: Pritt was a teacher and the daughter of a coal miner, you see; she was a *worker* and she beat an *owner*. The '96 election and Manchin's later political ascent prove less about the left's potential in West Virginia than they reflect a common truth: bosses cheat to win.

This fact of life has been proven once again by Richard Ojeda, a West Virginia state senator who recently parlayed an unsuccessful bid for Congress into the beginnings of a presidential campaign, one of the first in the Democrat's 2020 field. Ojeda became an outspoken defender of teachers during their strike, and he garnered a lot of attention by reminding West Virginians of their history. "Teachers are joining a fight for the soul and spirit of West Virginia that started hundreds of years ago," he wrote in a long Twitter thread. "Hundreds of years ago, investors came to West Virginia and purchased most of the land for all but nothing. Even today, you will be hard pressed to find many people in WV who own their mineral rights."

His logic is simple: West Virginia's workers, whether coal miners or teachers, have never benefitted from the state's natural wealth due to greedy corporations and the politicians they buy. Ojeda's plainspoken, often angry distillations of the state's woes, and his tender attention to the plight of teachers, grabbed national attention. He was featured as an example of someone who could turn the tide for Democrats in "Middle American places where their party used to prevail, but has been decimated in the Trump era," as journalist Trip Gabriel suggests.

Ojeda's willingness to stoke rather than soothe the growing militancy of West Virginia's rank and file undermines the claim that West Virginia is doomed to centrist Democrats. His success should not be underestimated. But, like Manchin and current governor Jim Justice (a lifelong Republican who got elected as a Democrat only to defect back to the GOP), Ojeda also has a history of straying to the right. Ojeda voted for Trump (a decision he now says he regrets) and has said he might switch teams. Moreover, Ojeda's evasiveness about the appeal of Trump's racism among West Virginia voters makes it difficult to embrace him as a unifying leader of the rural left. Indeed, I am reminded of former Republican senator James E. Watson's response to Wendell Willkie when Willkie switched parties, from Democrat to Republican, to run for president in 1940. Asked by Willkie if he believed in the power of conversion, a skeptical Watson replied, "It's all right if the town whore joins the church, but they don't let her lead the choir the first night."

Ojeda's presidential bid also sends a strong signal that he approves of the media ecosystem that has branded him a novelty. That novelty is dangerous, not least because pundits and political reporters are eager to propagate the idea that to win in places such as West Virginia—and now, it seems, the nation at large—the left needs its own version of Trump: a brash populist, prone to macho posturing, with a no-bullshit persona and little time for the rules or party politics.

But the more Ojeda's star rises, the further it departs from its grassroots origins, including a labor movement that is 75 percent women. In a story announcing Ojeda's presidential bid, for example, the Intercept initially ran a headline that referred to him as the leader of the teachers' strike—not just a vocal supporter. This mistake, although minor and originating from editors and not Ojeda's team, is reminiscent of the historical erasure of women's political work in Appalachia, particularly of women such as Pritt who are further to the left of West Virginia's Democratic establishment. More recently, there was Paula Jean Swearengin, Manchin's primary challenger and environmental activist, and Talley Sergent, who performed just as well as Ojeda in the recent midterms without generating a fraction of his national interest.

THE 2016 ELECTION still looms over us. But if all you know—or care to know—about Appalachia are election results, then you miss the potential for change. It might feel natural to assume, for example, that the region is doomed to elect conservative leadership. It might seem smart to point at the "D" beside Joe Manchin's name and think, "It's better than nothing." There might be some fleeting concession to political diversity, but in a way that makes it the exception rather than the rule—a spot of blue in Trump Country.

If you believe this, then you might find these examples thin: worthy of individual commendation, but not indicative of the potential for radical change. But where you might look for change, I look for continuity, and it is there that I find the future of the left.

It matters that workers are rising up, and it matters that women are leading. It matters that the fight against extractive capitalism is fiercer than ever. And for all of these actions, it matters that the reasoning is not simply, "this is what is right," but also, "this is what we do." That reclamation of identity is powerful. Here, the greatest possible rebuke to the forces that gave us Trump will not be people outside of the region writing sneering columns, and it likely will not start with electoral politics. It will come from ordinary people who turn to their neighbors, relatives, and friends and ask, through their actions, "Which side are you on?"

"Listen to today's socialists," political scientist Corey Robin writes,

> and you'll hear less the language of poverty than of power. Mr. Sanders invokes the 1 percent. Ms. Ocasio-Cortez speaks to and for the 'working class'—not 'working people' or 'working families,' homey phrases meant to soften and soothe. The 1 percent and the working class are not economic descriptors. They're political accusations. They split society in two, declaring one side the illegitimate ruler of the other; one side the taker of the other's freedom, power and promise.

This is a language the left knows well in Appalachia and many other rural communities. "The socialist argument against capitalism," Robin says, "isn't that it makes us poor. It's that it makes us unfree." Indeed, the state motto of West Virginia is *montani semper liberi*: mountaineers are always free. It was adopted in 1863 to mark West Virginia's secession from Virginia, a victory that meant these new citizens would not fight a rich man's war.

There are moments when that freedom feels, to me, unearned. How can one look at our economic conditions and who we have helped elect and claim freedom? But then I imagine the power of people who face their suffering head on and still say, "I am free." There is no need to visit the future to see the truth in that. There is freedom in fighting old battles because it means that the other side has not won.

Catte

Why Institutions Drive Change

Michael Kazin

HISTORICAL MEMORY is both a powerful and perilous thing. It can inspire you to emulate the great deeds of your forerunners. But it can also offer the balm of false comfort when what you really need is a stiff shot of reality.

U.S. leftists may have a particular weakness for romanticizing our predecessors. After all, most of our political victories have been fleeting and ambiguous, our heroes and heroines either little known (Wendell Phillips, Florence Kelley) or scrubbed of their radical thoughts and ambitions (Martin Luther King, Jr.) by the guardians of civil religion.

In her passionate essay about the rebellion she sees brewing in Appalachia today, Elizabeth Catte declares, "the way forward for the left, in my world, is through the past." Her historical examples certainly played a pivotal role in the organized uprisings that won a measure of economic and political power for working-class people during the first half of the twentieth century. The unionists who waged bitter, often successful coal strikes in places such as Harlan County, Kentucky, and Mingo County, West Virginia, helped build the United Mine Workers of America (UMWA) into a mighty force in what used to be one of the nation's key industries. In 1946 the union compelled mine owners to finance a system of health clinics and pensions that were the envy of other industrial workers. UMWA president John L. Lewis also founded the Congress of Industrial Organizations, whose member unions went on to organize the biggest manufacturing firms in the nation—and injected the New Deal and the Democratic Party with a healthy dose of class consciousness.

But that spirit and strength depended on a political economy that no longer exists. The coal regime was the latest—probably the last— chapter in an often painful narrative of natural resource extraction from the mountains and high plateaus of Appalachia. It yielded great wealth for a few and poverty for the many, only occasionally mingled with decent jobs and benefits. In his brilliant book *Ramp Hollow* (2017), Steven Stoll provides a history of what he calls "the ordeal of Appalachia." In contemporary West Virginia, he writes, politicians "continually menace" residents "with a false choice: health or jobs." Due to the lower price of natural gas, as well as necessary environmental protections and falling demand from other nations, most coal-mining jobs have disappeared. Despite Donald Trump's big promises, they are never coming back.

So the UMWA shrank. Nationally it has contracted to some 70,000 members (active and retired), down from a half million at its peak under Lewis. The remaining mine operators, with Republican support, slashed stipends for retirees. West Virginians became, once again, among the poorest and sickest people in the United States.

At the same time, voters there and across Appalachia recoiled from the cosmopolitan liberals who had become the national voices of a party once ruled by Franlin D. Roosevelt and John F. Kennedy, Jr. Many gave credence instead to the siren songs broadcast by Fox News and delivered from the pulpits of their churches. As a result, Republican presidential nominees have carried both West Virginia and Kentucky handily since 2000, racking up huge margins in old coal counties. In 2016 Trump got 83 percent of the vote in Mingo County. Across the border in Harlan County, he pulled down a whopping 85 percent. There may still be "no neutrals there," as Florence Reece put it, but there cannot be more than a hollow or two of union-loving progressives, either.

Catte knows this grim story. Still she claims that a different future is not only possible but is already underway. What is her evidence that a mass of Appalachians are busy creating a new movement to emulate the labor-centered one that disappeared decades ago? She mentions a brave Waffle House worker from Virginia who campaigned for a $15 minimum wage, a teacher from Mingo who took part in the recent strike against the atrocious conditions in the state's schools, and a Democrat with a long Army career who ran an aggressively populist campaign for Congress in southern West Virginia last year but lost by thirteen points. We should root for these people. But, taken together, their efforts, even if multiplied tenfold, add up to a pretty meager resistance either to big business or the Trumpified GOP.

Unfortunately, Catte's evocation of how such folks are using history to stoke their activism leans almost entirely on quotes about family members who took part in coal strikes thirty years ago. That is a kind of "continuity," but since it is based on struggles that failed to change—much less revive—a declining industry, I doubt it is one that has the potential to "drive forward the present movements," as she puts it with more hope than confidence.

FOR LEFTISTS to appeal to people in Appalachia and other parts of white rural America, they will have to talk in concrete terms about how to create secure jobs at decent pay, with the kind of benefits the UMWA once provided to its members. Rhapsodizing about the glory days of unionism will not convince those who, through no fault of their own, have missed out on the high-tech economic boom that has made metropolitan hubs such as the Bay Area, Seattle, and Northern Virginia so prosperous, if unequally so, and that has attracted a multinational workforce that votes reliably Democratic.

While Appalachians wait for that revival to occur, there is one lesson from the bygone days of working-class power that might help them stir or at least imagine a second coming of the left: build institutions that teach people how the world works and how they might change it. The UMWA accumulated clout at work and politics not just because it organized the men who did hard, essential work. The union also educated its members and their friends and families about who held power in the economy, which politicians were on their side and which were lying to them, and how to deploy a repertoire of tactics from shutting down a mine to effectively lobbying a state legislature.

A large and potent left once thrived in other parts of rural and small-town America too. Tenant farmers in central Texas railed at corporate moguls and the politicians who did their bidding. Their resentment and hope for redemption turned them first into backers of the Populists, then of William Jennings Bryan, and then of the New Deal. In 1908 and again in 1912, a higher percentage of Oklahomans voted for the Socialist Party of Eugene Debs, who hailed from a railroad town in Indiana, than did the citizens of any other state besides Nevada. Unless and until rural people build equivalents of the bygone farmers' alliances, feisty insurgent newspapers, and craft unions of old, progressive Democrats today will be vulnerable to the charge that they are, or represent, culturally alien outsiders who threaten the values of "real Americans."

Stories about the past can warm the heart and embolden the imagination. But only institutions can begin to turn those dreams into power.

Kazin

Class Matters

Nancy Isenberg

ELIZABETH CATTE IS RIGHT that the media treated Donald Trump voters as a group needing to be explained. Pundits in search of a grand narrative found it in J. D. Vance's *Hillbilly Elegy* (2016). It helped that Vance had escaped the great rural unknown to attend Yale Law School and work in Silicon Valley. His memoir told the story of a dysfunctional family that could, some thought, stand for all of Appalachia—and by extension, all white working-class Americans trapped in dying cities or rural wastelands.

Vance's book received praise from both liberals and conservatives. For Republicans, it was a tale of self-reinvention and self-reliance: he refused to blame the state for his family's troubles and claimed, in the end, that his mother had only herself to blame for her addiction. The message was simple: pull yourself up by your own bootstraps. For liberals and progressives, Vance painted a picture of an alien world of rural whites trapped in false consciousness and a spiral of self-destructive tendencies. Theirs was a broken culture that required diagnosis. By turning Hillbilly Central—West Virginia—into Trump Country, the media did what they usually do, reducing the complicated history of a diverse region into a snappy sound bite of white rage.

Throughout the 2016 election season, the faces of that Trumpian anger were charged-up men in trucker hats and women in T-shirts chanting "Lock her up." Trump tapped into class resentments, bragging about his love for the "poorly educated." If Washington elites were privileged insiders, mired in the tax-producing swamp, his supporters were the real Americans, the hardworking majority.

The stark irony of it all was that an avowed multi-billionaire had stepped down from his New York City penthouse and found a formula that enabled him to forge a bond with the masses. The national media coded Trump's base as the working-class "other"—white trash, rednecks, hillbillies. It was supposed to be an identity that skimmed, even ignored, real economic issues—the embodiment of vague "economic anxiety" (a diagnosis laden with psychological baggage), as if the wound was self-inflicted and detached from the nation's long history of class oppression and division.

The 2018 midterm elections suffer from simplistic characterization as well. The whopping victory for Democrats in Congress foretells a new "year of the woman," we have been told. True, a diverse array of female faces were sworn in, and Alexandria Ocasio-Cortez of the Bronx has become the poster child for a new socialist-inspired movement. Yet much of the media seems incapable of seeing that gender and class are not exclusive categories. As Catte shows us, Ocasio-Cortez's socialist message resonates beyond hip cafés in Brooklyn; socialism has a long history of influence in the South as well. Though not a member of the Socialist Party of America, Louisiana's Huey Long, Jr., was not a good-old-boy populist who came out of nowhere. His "Share Our Wealth" program of the Great Depression echoed what left-leaning southern progressives were promoting in the 1930s. But Americans have short historical memories. Catte reminds us that West Virginia, like rural America's political past more broadly, is not to be boiled down to the simple catch-all of a "red state" mentality.

I WOULD ADD two troubling trends to Catte's diagnosis of the present. One is the rejection of class as a factor in Trump's election. Dan Balz and Jennifer Rubin of the *Washington Post*, for example, have discounted assessments that say economic concerns mattered. Referring to the argument made by political scientists John Sides, Michael Tesler, and Lynn Vavreck in their book *Identity Crisis* (2018), Balz approvingly declared this past September that "race, religion, gender and ethnicity" were the "driving forces" behind people's votes, especially white voters. Class identity doesn't even make the list, as if class doesn't produce an identity.

The other trend since the midterms has been the desire to make Ocasio-Cortez the most visible symbol of the blue wave. The cult of celebrity is built into U.S. political DNA—the one thing that George Washington, Eugene Debs, and Martin Luther King, Jr., all had in common was star power—and it is not easily defanged. But I would like to see other progressive women get media attention, such as Sharice Davids, the LGBT Native American attorney from Kansas. There is also a lesson to be learned from Catte's example of Virginian organizer Nic Smith, whose appearance in viral videos has helped to revive the important New Deal and Great Society message that poor people in rural coal country are suffering because they do not own the land. As he said in an interview in 2017, the way to beat Trump is to follow the example of JFK and Lyndon B. Johnson. "They went to coal towns and said, 'We're here. And we give a shit.'" But where is Smith now? The media have moved on to new faces.

I also have less faith than Catte about the ability of rural voters to revive their socialist heritage by drawing on family and generational memories. Memory is not immune to distortion. West Virginian and former Army major Richard Ojeda demonstrates this trap, and Catte is right to point out the limitations in seeing him as a "novelty" candidate. He is appealing precisely because he fits the conventional masculine image of a rural populist and labor leader. Like all Americans, West Virginians are still conditioned by gendered expectations, and the blue wave did not wash them away. Mother Jones would not stand a chance of getting elected today.

As a historian, I cannot honestly romanticize any past political tradition. Political theorist Corey Robin may say that among today's socialists "you'll hear less the language of poverty than of power," but U.S. party politics has always indulged in exaggerated rhetoric. It is not enough to get voters excited. I would like to see socialists, Democrats, and the media do more to discuss the difficulty of translating practical social problems into viable political solutions. Achieving political change through legislative victories is an arduous process, and optimistic messaging is not enough.

Socialism can and should inform this work, but we must do more than look to the past. Socialism's influence has not been uniform, after all, even in the United States; Huey Long's vision is vastly different from that of feminist Charlotte Perkins Gilman. Democratic candidates must address how *our* society distributes resources unequally when it comes to education, health care, safe neighborhoods, economic incentives, tax breaks, infrastructure, and the environment. A single socialist game plan may not easily bridge the urban and rural: Bernie Sanders (who represents predominantly rural Vermont) and Kirsten Gillibrand (when she represented a rural New York district) were outliers among Democrats on gun control. Sanders's views on race have also remained trapped in 1960s rhetoric about the "ghetto" (which helps to explain why most black women voters sided with Hillary Clinton). My point is that socialism has some serious blind spots in addressing today's issues that cannot draw on history for answers.

We don't just need new faces; we need more honesty about how class shapes the lives of everyone. Democrats must confront class blindness with the same tough talk they demand in exposing racial or gender prejudice.

The Last Steep Ascent

Bob Moser

I GREW UP QUEER in a white working-class North Carolina clan during the 1970s, that moment in history when the backlash to civil rights and feminism and unionism was beginning to gather itself into the regressive forces that became Republicanism and Democratic Clintonism. If I learned anything from the history I would huddle in my room and read obsessively—and later, from the black friends and boyfriends I could not bring home, and from my reporting on the politics and progressive movements of the South—it was this: our only hope for freedom was a radically different future.

Emphasis on "radically." Not the kinds of victories that won legal rights—which labor, African Americans, and women had sort of done, and LBGTQ folks now, sort of, have too. No: the only hope, thin though it was, was in attaining real power. There could be no real progress without destroying the white corporatocracy that passes for democracy—the ultimate victory that the left's brave and battered standard-bearers, among them those who figure in Elizabeth Catte's moving essay, have so persistently failed to achieve.

Many of the progressives I have reported on and protested with over the years dearly love to celebrate old glories by singing old songs and lauding old martyrs, as if the breakthroughs they had occasionally inspired were not profoundly depressing for being exceptions to the rule of defeat. These folks were my allies, but how sad and delusional they could be—firing up the candles for another midnight vigil, passing the talking sticks, murmuring another round of "This Little Light of Mine," and otherwise luxuriating in righteous failure. The left has been as addicted to its various Lost Causes as the neo-Confederates I also knew and wrote about. And that, I came increasingly to believe, was why we never stopped losing.

Don't get me wrong: I am not immune to sentimentality. Like most on the left, I fell prey to it all over again when Barack Obama burst forth in all his symbolic and rhetorical glory in 2008. Ten years later I was briefly falling for Beto O'Rourke, another charismatic agent of airy hope and change whose actual politics, when you finally take a hard look, are every bit as corporate-friendly and unchallenging as Obama's proved to be.

Thankfully, though, while Beto was peddling the same old snake oil, genuine change-makers were rising down South in the wake of Obama's betrayals and Donald Trump's ascendancy. Hardly anyone took notice at first, when in 2017 the largest cities in Alabama and Missis-sippi elected young black mayors backed by Bernie Sanders—Randall Woodfin in Birmingham and Chokwe Antar Lumumba in Jackson, the latter of whom won (can you believe it?) on a promise to build "the most radical city on the planet." Little ink was spilt, as well, when an all-black slate of officials and judges, including Black Lives Matter activist khalid kamau, took power in South Fulton, Georgia—the city they had worked to incorporate by uniting a string of majority-black, overwhelmingly poor counties in the Atlanta suburbs.

A few eyebrows did lift in 2017 when Virginia's first transgender and Asian American and Latina delegates ousted right-wingers in the statehouse, and thirty-eight-year-old Justin Fairfax became the state's second African American lieutenant governor and chief executive in waiting. Fairfax had an ancestor's freedom papers in his pocket when he was sworn in. "They tell people to get in line, but the problem is there is no line," he has said. "This generation has figured it out. People will say that your future is bright. . . . I say your future is now. . . . It's a very different paradigm."

Different indeed. The strongest signal that something truly new has been emerging down South came with Alabama's election of a white centrist, Doug Jones, to the U.S. Senate in a runoff against Bible-thumping, Confederacy-loving Roy Moore. How in God's name had *that* gone down? The sexual assault allegations helped, but the real answer was a grassroots upswelling of black activists who saw their opening and aimed at far more than electing Democrats.

DeJuana Thompson is one of them. When she organized for Obama in North Carolina and Florida, she saw firsthand what electoral victories could be achieved when Democrats stopped obsessing over winning back Reagan Democrats and focused instead on galvanizing the Sunbelt South's emerging majority of non-whites and liberal millennials. She later saw how little those victories meant in terms of tangible progress. The group she founded in the waning days of Obama's presidency, Woke Vote, is among those that now aim at more radical change. The message to Democrats, she told me last summer, has to be clear as day: "We're not waiting for you anymore."

You will hear the same thing from insurgent politicians such as Lumumba and kamau, and from the whole raft of young southern organizers bent on liberation. Black, Latino, Asian, and millennial turnout shattered state records in 2017, and national ones again in 2018, thanks to the work of fellow organizers LaTosha Brown and Cliff Albright of the Alabama-based Black Voters Matter, and Cristina Tzintzun Ramirez of Texas's Jolt Initiative, which helped to seed a similar insurgency among young Latinos. Yes, Stacey Abrams and Andrew Gillum did lose—if just barely—their bids for the governorships of Georgia and Florida. But we should take note of the success they found by throwing away the old Democratic playbook of Republicanism Lite, running on ambitious left-wing platforms that have more in common with Alexandria Ocasio-Cortez than Bill Clinton, and building their campaigns on empowerment. Rather than appeal to white people's "better angels," they called, as Thompson has said, for "liberating ourselves through our voting power." And they met their defeats with defiance.

"Democrats in the South have to reject the notion that our geography requires that politicians soften our commitment to equality and opportunity," Abrams said during her campaign. And this time it was not mere rhetoric. Like the organizers who damn near overcame rampant voter suppression to lift her and Gillum and O'Rourke to victories in races Democrats had not won for two decades or more, Abrams does not indulge in rose-colored recollections of Martin Luther King, Jr., and other progressive heroes of the past. She knows what King, the *real* King, knew: that the freedoms won back then by the Civil Rights and Voting Rights Acts were important but also completely insufficient. Rather than bask in the glow of those victories, King recognized their limitations: "The prohibition of barbaric behavior, while beneficial to the victim," he wrote, "does not constitute the attainment of equality or freedom." In 1966 he called the more consequential battle to come "the last steep ascent."

The man who wrote these words was not the "dreamer" we have chosen to recall and celebrate, but a clear-eyed realist. I imagine he would be saddened but unsurprised to see that we have failed to scale that last steep ascent. But he might have taken heart at the rise of a new generation of radical realists who eschew nostalgia and have their eyes set firmly on the future—one that will not be built on "fighting old battles" and looking for "continuity," in Catte's words, but on the collective, determined accrual of power. "We are going to rescue ourselves," Lumumba says. That is not a backward-looking expression of hope; it is a revolutionary promise, entirely in the future tense.

Legacies of Resistance

Ash-Lee Woodard Henderson

SINCE 2016 the number of people who want to chat with me about rural communities, particularly in the South, has dramatically increased. Popular questions include how the Highlander Research and Education Center and our southern organizational partners prioritize, engage, and communicate our missions in this specific "political moment."

To think of this as a specific moment, however, could not be further from the truth. This is a political present, and it is not disconnected from a dialectically incredible and troubling past. The special magic of being connected to the southern freedom and black liberation movements is that we attempt to practice the liberation in the present that we had hoped to live into. Elizabeth Catte's reflection shows how that connectedness to our radical legacies of resistance informs so much of what is possible in Appalachia and across the South.

These conversations about strategies and ideologies are not new for people of good will in our region. Chattanooga, Tennessee, for example, threatened to secede from the state because of the city's anti-slavery, pro-Union politics. By the 1930s, it was a headquarters for the Communist

Party. Examples similar to Chattanooga are available across the South; various cities and towns have influenced the tactics and strategies that we use today, and they feed people's hope in the idea that things can be different and better.

We are building on these past successes, even as they have been the target of right-wing, white supremacist regressive policies. Many of us have learned to not be distracted by false narratives about our region. We have learned to stand in our truth instead of just raging on social media, and when articles call us names that are not ours (e.g., backward, people who don't vote their best interests, Trump Country) we put in the work to prove who we are. The brilliant black feminist Alexis Pauline Gumbs once described a prophetic dream that Harriet Tubman had in 1862, after which Tubman proclaimed that "My people are free." Gumbs said, "If Harriet Tubman could talk about freedom in the present tense while posters around the country marked her as a 'fugitive slave' then we can robustly and specifically speak freedom in this moment when anti-blackness and state violence ring off the sidewalks, pools, patios, and streets of this land."

We continue to speak freedom, and, as a result of our efforts, we have shifted judge seats; we elected district attorneys who are accountable to our people; and we built people power across Appalachia and the South. Our people have rallied behind Andrew Gillum and Stacey Abrams. In Florida we passed Amendment 4, restoring voting rights to some convicted felons.

I mention these lessons because they create a question for our colleagues who live outside the region. Which stories are you amplifying? Where is your camera pointed? Is it pointed here, at rural communities and inner cities, because we have solutions? Or because you have decided we are the problem?

The Democratic Party refuses to invest in the most populous geographic region in the country. Too many organized liberal, progressive, and left forces concede Appalachia and the South to white nationalists, white supremacists, and the extremist Tea Party. Now, as a result, an entire region has been reduced to "Trump Country." We are blamed for Trump based on voting margins, even though those numbers do not take into account the many people who did not vote and, even more importantly, could not vote.

Our reality is that neoliberal politicians (at best) and outright conservatives in Democrat's clothing (at worst) have not been good for our people. Chattanooga is a case in point. With depressingly low voter turnout, the city elected Andy Berke, the Democratic nominee, as mayor in 2013. The city has since been home to one of the fastest-gentrifying zip-codes in the country, even as it also has the highest concentration of low-wage workers in the country. Under the guise of progressivism, liberalism, and innovation, Chattanooga has become what Concerned Citizens for Justice deemed "The Perfect Storm of Inequality."

Catte is spot on with her assessment that the way forward is through the past—a concept the Asante people of Ghana call "Sankofa." It is why the Highlander Center is busy developing the next cohort of young southern freedom fighters and supporting them as they deepen transformative relationships with each other and with our movement elders. It is why we have brought rural white Appalachian factory workers together with Mexican factory workers, both exploited by the same capitalist bosses who tried to pit them against each other. The dictatorship of the elite and the wealthy, the corporations they run, the politicians they pay off, the systems of oppression that they uphold that harm our people have dismembered us from our radical legacies of resistance. It is our work to remember. And it is the job of allies outside the region to support, accompany, flank, and amplify those of us doing that work.

In the wake of the 2018 midterm elections, North Carolina organizer Julia Sendor said that white people—especially white women—need to ask the right questions to understand and dismantle whatever is blocking them from believing in and learning from rural Appalachian folks, particularly women of color. What is blocking people born and living outside the region from seeing vital truths as truths? What combination of assumptions, misinformation, lack of information, and even (and often) personal insecurities and fears are we all holding onto and spreading?

These questions are important because the one thing I know for sure is that as goes the South, so goes the nation. Folks in Appalachia are contending for power against white nationalists and white supremacists, and they are having a positive, power-building, lifesaving impact. The question is not *Is the work happening?* but *What would happen if folks committed to supporting and investing in it?*

Left Behind

Elaine C. Kamarck

ELIZABETH CATTE HAS PROVIDED a powerful corrective to conventional wisdom about rural America. Too many urban progressives view rural America as outside their "bubble." This is not only condescending but wrong. Like other Americans, rural Americans have always known their own interests, and rural progressives have a strong track record of organizing to protect them. Catte reminds us that the struggles against coal companies in Harlan, Kentucky, in the 1930s, and the struggles of teachers in West Virginia more recently, are all part of a powerful collective action narrative.

Her remembrances and her hopes of a rebirth of activism against "the plundering coal bosses," however, come up against a economic reality to which no one—from the center left to the far left—has any good answers: We know how to organize against the bosses, but what do you do when there are no more bosses?

If only rural America still had "plundering bosses." The twenty-first-century economic reality is much more difficult. We now face two United States: one with bosses and jobs, one with neither. A graph from a recent Brookings Institution paper on "left-behind places" illustrates the problem. In 2008, at the start of the great recession, employment rates did not vary by the size of the community. During the recession, of course, employment rates everywhere went down. But starting in 2013, when the recovery picked up enough to start creating jobs, something new and different happened: "Big, techy metros like San Francisco, Boston and New York with populations over 1 million have flourished, accounting for 72 percent of the nation's employment growth since the financial crisis."

This lopsided economic growth came as a surprise to many. Economists have always clung to the belief that a rising tide would lift all boats. One means of uplift, historically, has been that people were willing and able to move from places where there were no jobs to places where there were jobs. One of the most famous migrations, of course, happened more than a hundred years ago when southern blacks left the violence and poverty of the South to seek work in the new factories of the Northeast and Midwest. Economic migration today, by contrast, seems to have stalled. Low levels of digital skills in smaller communities, combined with skyrocketing housing prices in larger communities, stifle mobility. Moreover, many of the "left-behind" places lack the access to capital, transportation infrastructure, and broadband that could create homegrown economic opportunities.

West Virginia, one of the focuses of Catte's essay, has been especially hard hit by changes in the economy that have little to do with the bosses. With one industry dying and no other rising to fill the void, there is nothing tangible or immediate to organize around or fight against. Donald Trump's success was in rallying this helpless feeling into anger against outsiders and the financial or knowledge-economy elites. There are surely more productive ways to rally—demanding high-quality, free technical training at community colleges or money for small business startups, for example—but the crucial point remains that the bosses in today's economy are far removed from West Virginia; traditional forms of organizing are unlikely to even register on their radar. Consider the fact that even if the country or the world were to reverse course and start demanding more coal, thereby reviving West Virginia's main industry, it would not create more jobs. *Computerworld* reported that "in the next decade, the mining industry may lose more than half of its jobs to automation." Robots do not get black lung disease nor can they be organized.

Which is why, at the end of the day, Catte's nostalgia leaves us with no clear path. Neither Alexandria Ocasio-Cortez (whom Catte admires) nor Joe Manchin (whom she does not) have the answers to the dilemmas posed by the twenty-first-century economy. Ocasio-Cortez's Green New Deal, for example, is a wonderful initiative, but hardly new or paradigm-shifting. The federal government has been incentivizing green jobs since the early 1990s when it passed tax credits for renewable energy. Today federal tax credits for renewable energy account for 59 percent of all energy tax credits, more than double the credits for fossil fuels (25 percent). These credits have been invaluable in moving these industries—especially solar and wind—along, and they can always be increased or combined with an infrastructure program (which is the real virtue of the Green New Deal), but saving places that have been left behind will require a lot more than solar jobs.

Catte is at her most instructive not in the exhortation to organize against bosses and capital, but rather in her analysis of the activism and victory seen in the West Virginia teachers' strike. That victory did more than help to improve economic equality and dignity for teachers (a worthy cause). It also increased the chances for an economic future in West Virginia. An educated workforce is increasingly critical; without it, the economic future of West Virginia and other left-behind areas is bleak.

Politicians, then, not corporate bosses, are the most important powers to organize around today. And while it is fair to hope for a revolutionary who could rethink the system and save these communities in a single presidential term, it is also crucial to realize that there are no easy or short-term fixes for economic dislocation. Transfer payments—such as social security, disability insurance, Medicaid, and Medicare—have gone disproportionately to those areas that have been left behind, and they are the only efforts that consistently soften the blow and afford people a modicum of dignity.

Democrats, even conservative ones such as Manchin, have been on the front lines protecting a wide array of transfer payments. Manchin has been a staunch supporter of Medicaid and of Obamacare, voting repeatedly against Republican efforts to undo it. Democrats in the House and Senate have also held the line against the Paul Ryan austerity budgets, and they will continue to hold the line against Senate Majority Leader Mitch McConnell's desire to see the budget balanced through cuts to entitlement programs. So, yes, I understand why Catte wants West Virginia to do better than Manchin. But I sure hope she still voted for him.

Don't Blame Capitalism

Matt Stoller

ELIZABETH CATTE'S ESSAY on the left-wing tradition in Appalachia is not so much a political argument as a beautifully written reflection on an important cultural tradition of dissent. She is part of a trend of thinkers and organizers who have come to reject capitalism per se as an immoral system that must be overthrown.

I believe in market systems and commerce. But I get why a large chunk of younger people think capitalism is bad. Most have never seen anything but crappy political economy policy organized by hypocrites and bureaucrats. If you are in your twenties and thirties, every political leader and op-ed writer and labor organizer and academic you have heard has told you that Obamacare—with its nightmarish copays and coinsurance and weird extractive bureaucracy—is the best you can get. Confronted with the prospect of a life of dreary disappointment and climate change, any rational person told to clap louder is going to say, screw that, let's get rid of capitalism or whatever this thing is we have now.

I am slightly older, but my experience is not that different. I got into politics in 2002 as I watched immoral wars get peddled by lying journalists. I then worked as a staffer in Congress on a terrible set of bailout policies that ended up funneling wealth to greedy crooks and foreclosing on the middle class.

So I get it. But what is driving this impulse—whether in Appalachia or in Brooklyn—is not dislike of capitalism, per se, or disdain for markets. We have had market systems that incorporate slavery, and we have had market systems that promote freedom. Lumping them all under the name "capitalism" is conceptually lazy. Markets and finance are engineered social systems that can be used to promote freedom or undermine it, and throughout our history they have done both. A farmer's market and a derivatives market are not the same thing, even if they both use the word market.

The political problem we face today—in Appalachia and elsewhere—boils down not to a debate over socialism versus capitalism, but to something simpler. Obama was a really important president at a pivotal moment in history, when a financial crisis gave him wide latitude to restructure our social obligations. And he screwed it up. Some 10 million foreclosures and no Wall Street felons. There are a lot of other ways he restructured society to make it less free and more unequal. For example, because of the way bailouts were structured, black-owned banks were a tenth as likely to get bailout money as other banks. Obama's antitrust officials allowed mergers in telecoms, pharmaceuticals, airlines, and tech platforms, concentrating power in radical ways. Obama negotiated a bill to hand over Puerto Rico to hedge funds. And what did Obama do about opioids in rural America? A friend of mine in the administration told me that when the White House finally noticed the AIDS-level epidemic death toll, the suggestions proffered were . . . roundtables. She might have been exaggerating, but not by much.

The gist is that Obama reorganized our markets to push wealth and power upward, and to subvert our liberties. It is a painful story. It is not, however, a complex one. In 2008 we thought we were electing Franklin Delano Roosevelt, but really we elected Herbert Hoover.

Most centrists and left-wingers are embarrassed to entertain this simple tale. So they spin complex stores involving fancy theories. Instead of just saying, "Obama harmed tens of millions and induced a loss of faith in democracy, paving the way for Trump," they make excuses. There are a host of centrist ones, but the left-wing excuse is to offer the Grand Philosophical Explanation of Why Capitalism Is Bad. After all, if it is the system that is at fault, then no one has to admit they should not have celebrated this guy in 2008, or that the people who told them to celebrate him for eight years were incompetent frauds. Everyone gets to protect his or her ego.

Even today you cannot get a single elected left-wing politician to say that Obama was a bad president. Think about that. We cannot have an honest discussion of what it meant to use power when Democrats were in charge, so the language of dissension is polluted with incoherent nonsense. All the grand philosophical musing and Democratic Socialists of America study groups do not matter when not a single elected official outside the Republican Party can make the simple, obvious point that Obama's policies straight up made things worse.

This was not some capitalist plot. There was a lot of dissent within the Democratic Party about whether it was a good idea to do what Obama did. I was part of a network of people who tried to fight against the foreclosure nonsense and opposed Obama's handing Puerto Rico over to hedge funds. We lost. And the people who made public explanations about these fights lied to cover up for Obama's bad choices. They lied because some of them are frauds, but also because it was painful not to; Democratic voters and many left-wing voters were and still are deeply hostile to any criticism of Obama. He is beloved; according to Gallup polling, 95 percent of Democrats have a favorable view of him. To the extent there is skepticism, it is framed in ways that avoid admitting that his actions systemically ruined millions of lives.

It is not hard to get why people in Appalachia are mad. It is not cultural longing for some long-lost past. Hospitals are closing and it is hard to find a good doctor. Broadband is bad. It is hard to get anywhere because there are few flights, and there are few trains to ship crops and goods. Drug addiction is rampant. You can't get a bank loan. Local media is dead because ad revenue has been pulled to Google and Facebook, and Fox News has filled the void. People are ringed by financiers and monopolists and don't like it.

This is not a problem of capitalism. In 1980 these were politically healthier communities, and not because Americans were singing the "Internationale." We had a set of policies that engineered markets to enable freedom instead of subvert it. Our trains and airlines were regulated to ensure people could get to and from these regions. Our health care system was smaller and less corrupt. Our banks and farms were smaller and more local, so people had access to credit and nutritious food. Walmart and Dollar Tree did not destroy every independent shop, because there were laws restraining chain store expansion and protecting local communities and local manufacturing.

We can achieve a free society if we organize regulated competitive markets, and that is what we should do. But before some grand ideological debate takes place, Democrats and the left have to just admit Obama screwed up in dangerous ways, and the institutions of the left and Democrats covered up for him.

The Radical History of Appalachian Women Activists

Jessica Wilkerson

ELIZABETH CATTE NOTES that women's grassroots leadership has been fundamental to progressive movements in Appalachia. She offers illustrations that range from Florence Reece penning the ballad "Which Side Are You On?" during a 1930s labor struggle, to teachers in Mingo County, West Virginia, leading the state's teacher strike.

I agree that we can learn from the history of women's activism in Appalachia—and women's relationship to extractive capitalism—as we accept Catte's call to move forward "through the past." In response, I would like to provide further context by discussing another crucial moment in the history of women's grassroots organizing, when women drew upon the zeitgeist of the 1960s and took advantage of Great Society legislation, most notably during and after the War on Poverty.

The consensus in Appalachia is that the Johnson administration's War on Poverty was a failed liberal experiment. That characterization forgets how War on Poverty initiatives were vigorously opposed by local elites who served the interests of the coal industry, calling for investigations of antipoverty programs as a way to undercut their credibility.

It also erases from memory the grassroots organizing of women. They worked in alliance with activists from across the region and country to implement and shape the most progressive legislation since the New Deal. They also envisioned a new economic future for Appalachia. Although the War on Poverty was cut short by the fracturing of the Democratic Party and conservative backlash, these women's vision—for redistributive economic policies and a collective sense of fairness—remain an extraordinary source of political inspiration.

Edith Easterling was one such organizer. In 1964 she was a resident of one of the heaviest coal-producing counties in eastern Kentucky and the spouse of a disabled coal miner. Theirs was a pro-union household, and Easterling was politically active in her rural white community. She participated in the local PTA, making sure that resources were distributed to poor rural kids, and she campaigned for local Republican candidates, the "mountain Republicans" Catte describes in her essay. Easterling's loyalties were not to a political party, however, but to her family and neighbors. So, when Lyndon Johnson and other Democrats promised federal aid to impoverished areas of Appalachia, Easterling wanted to know more.

Easterling took seriously the Economic Opportunity Act (1964), the signature legislation of the War on Poverty, when it called for "maximum feasible participation" of the poor in antipoverty programs. She soon joined forces with the Appalachian Volunteers, a federally funded service program led by young, white, college-educated men and women. The majority were not from Appalachian communities, but there were exceptions: Easterling's daughter Sue Ella was among the first to join the organization, and she introduced her mother to federal antipoverty workers. Easterling soon became a paid organizer.

Starting in 1966, Easterling worked with antipoverty activist Joe Mulloy, a young Kentucky native inspired to join the Appalachian

Volunteers because he wanted to be a part of progressive change in the South. Together they began organizing poor and working-class people in the Marrowbone community in Pike County, Kentucky. They opened a community center and library, where they ran workshops on a range of topics, from welfare rights to Appalachian labor history. They also joined the burgeoning movement against strip mining and helped expand it, exposing the costs of extractive capitalism on working-class communities. Easterling soon became part of a network of progressive activists in the South, and she and her husband opened their home to civil rights activists who traveled to the region, hoping to build alliances between black and white workers.

Easterling's activism eventually earned her a spot on the board of the Highlander Research and Education Center in Tennessee, a nationally known training school for progressive activists. It also put her, along with Mulloy and others, in the crosshairs of local officials who claimed outsider activists were plotting a takeover of eastern Kentucky.

The backlash extended into 1968, when the Kentucky state legislature created the Kentucky Un-American Activities Committee (KUAC). Initially formed to stymie civil rights and Black Power activism in Louisville, the committee soon targeted majority-white antipoverty organizations as well. KUAC called Easterling to testify and questioned her about her relationships with civil rights activists and so-called "subversives." Easterling defended her allies and refuted the claims of local elites, who, she argued, cared more about lining their pockets than expanding democracy.

The following year, Easterling sent a friend her appraisal of the state of activism in Appalachia: "Interest has aroused, stones unturned, dander up, fever risen, and people speaking out on what they [believe]." She thought the movement would weather the backlash, and it did for a time. Well into the 1970s, Appalachian activists—many of them women—continued organizing. Their focus became redistributive policies, including a guaranteed income, access to health care, and welfare as a right of citizenship.

No component of the movement better captured the philosophy and optimism of the period than did the mixed-sex, interracial, regional organizing around welfare rights, a campaign led nationally by black women and fueled by the Poor People's Campaign that grew out of Martin Luther King, Jr's civil rights activism. For instance, the bylaws of the Eastern Kentucky Welfare Rights Organization, made up primarily of white welfare recipients and retired and disabled miners, stated its guiding principle that "the wealth created by the resources of this great country should be directed to give an adequate income to all its citizens."

West Virginia organizer Shelva Thompson saw welfare rights as a way to build solidarity among working people. "There's not just one group of oppressed people in the mountains," she argued. "We've got miners, we've got welfare recipients, we've got blacks, Indians. . . . But the government until now has kept those groups fighting each other till they never had time to fight the oppressor. And hopefully Appalachia is waking up."

As Robin D. G. Kelley has written previously in *Boston Review*: "We cannot change this country without winning over some portion of white working people, and I am not talking about gaining votes for the Democratic Party. I am talking about opening a path to freeing white people from the prison house of whiteness." Poor people's movements of the 1960s and '70s provide a blueprint for how to once again expose whiteness as "a foundational myth for the birth and consolidation of capitalism."

Thompson called for class solidarity, and she and others also centered women's unpaid caregiving labor in the coalfields. If they and their children were hungry, it was because capitalists owned the land

and paid workers low wages. If they were sick, it was because industry denied them health care and polluted the environment. If they lacked decent schools, it was because tax policies favored corporate landowners over working-class communities. When the economy failed working people, they argued, women were left to pick up the pieces, caring for children, disabled adults, and the elderly.

By the late 1970s, the most progressive movements in Appalachia hobbled under a racist and sexist backlash that targeted welfare recipients. At the same time, the Democratic Party failed to follow the lead of welfare rights activists, whose economic proposals centered caregivers and made connections between welfare rights and worker rights. Appalachian women's activism became muted. We must endeavor to recall and take inspiration from its radical spirit. If a working people's movement in Appalachia seems impossible today, it is because we have failed to remember how and why they organized in the past.

Queer in Rural America

Hugh Ryan

THERE IS A FAUX TRUISM in the United States that queer folks do not fare well in the countryside. As Elizabeth Catte says, "Rural spaces are often thought of as places absent of things, from people of color to modern amenities to radical politics." LGBTQ people could easily be added to that list. This sense of absence is continually reinforced by the media, which only talk about rural queerness in the context of high-profile murders or low-life government officials. Like many of the videos in Dan Savage's It Gets Better project, this oversight suggests that the route to happiness for gay folks leads inexorably away from the countryside. However, even if every queer person wanted to leave rural life behind, many would not have the option, and any movement that is bedrocked on leaving behind whole swaths of the country does not deserve the name "queer liberation."

In recent years there has been an explosion in rural queer organizing. In 2014 the National Center for Lesbian Rights—working with the U.S. Department of Agriculture—launched the #RuralPride campaign, which has since convened more than fifteen day-long events for queer people in places such as Lost River, West Virginia, and Wayne, Nebraska. Just this month, The Center for Lesbian and Gay Studies in New York City (one of the most prestigious centers for queer theory in the country) awarded its lifetime achievement award to Amber Hollibaugh, a communist sex radical writer and organizer who draws powerfully on her roots in "trailer park California."

Outside the world of conferences and academia, grassroots organizing is being done by groups such as Southerners on New Ground, a decades-old social justice organization that "envision[s] a multi-issue southern justice movement . . . in which LGBTQ people—poor and working class, immigrant, people of color, rural—take our rightful place as leaders shaping our region's legacy and future." Newer projects, such as Rachel Garringer's Country Queers oral history archive and the film *Forbidden: Undocumented and Queer in Rural America* (2016), seek to document the lives of rural queer people in their own words. YouTube is also fast becoming its own archive and networking tool for rural queers, providing a place to discuss everything from coming out in rural Virginia, to returning to the Navajo reservation in Crownpoint, New Mexico, to Arkansas's rural gay radio broadcasts of the 1990s. Paralleling this development, rural projects aimed at general audiences have begun to include queer content, such as the "LGBT in Appalachia" panels at the 2014 Appalachian Studies Association conference. Such mainstreamed initiatives may ultimately have the greatest success in reaching rural queer people, as recent writings have suggested that queer people who live in the countryside often value local community connection over connections based on sexuality, and thus are less likely to seek out specifically queer resources and networks.

These projects organize outward from the lived experiences of rural queer people. But there is also a growing movement of urban and suburban LGBTQ folks who dream of escaping *to*—not *from*—the countryside. These queer rural utopian projects take as their starting point a rejection of urbanism—which may or may not be based on any actual rural experience. They posit "the rural" as a panacea for the perceived problems of modernity, queer or otherwise: lack of community, loneliness, deracination, environmental devastation, personal disenchantment. Many seek to create queer rural oases that reject dominant U.S. social and sexual structures. Like similar religious communities in early America, many of these groups are to some degree millenarian, seeking an idyllic and self-reliant life in preparation for the inevitable and fast-approaching end time.

In their early days, queer rural utopian groups were often gender-essentializing, with lesbians creating wymyn's land collectives around the country, and gay men flocking to "faggot-only" Radical Faerie gatherings organized by communist luminary Harry Hay. The degree to which trans folks and people of color were welcome in these spaces varied from group to group, as was true in much of the mainstream lesbian and gay movements of the era.

As Catte points out, the countryside is often derided as a place where change comes slowly or not at all, but queer utopianism has evolved rapidly from those 1970s roots. Today, collectives such as Sojourners Land celebrate "Black Queer and Transgender Women," while a recent issue of the *Radical Faerie Digest* explored queer utopianism through the perspective of queers with disabilities. Although the popularity of such projects is hard to quantify, a quick search for the word "queer" in the directory of the Fellowship for Intentional Community brings up a whole host of other communes created by or embracing of queers.

Meaningful recognition of rural queer people will present some challenges for LGBTQ organizers. Perhaps top among them is the conundrum of how to work for the rights of people who may rarely (or never) acknowledge their membership in the rainbow tribe. So much of queer organizing in the past few decades has been predicated on ideas of visibility and coming out that it can be hard to think outside of that paradigm. Two of the movement's biggest recent victories—same-sex marriage and open service in the military—have nothing to offer for those who cannot or will not come out.

But there are creative solutions to these issues, which may in fact be more liberatory in the long run. Instead of just fighting to have our sexual relationships recognized on an equal basis with heterosexual relationships, we could question using sex as a marker for government benefits at all. What if we gave tax breaks to relationships that provided meaningful mutual support, regardless of the kind of relationship—sexual, familial, platonic—in question? Not only would this help queer people who are not out and may be living together as friends, roommates, or even cousins (all common strategies to hide queer relationships), but it would also make space for solidarity with groups working for the rights of multigenerational families, single people, immigrant communities, and working-class people.

Of course, not all rural queer people are closeted or even "discreet" (a term of art on gay dating sites). And I am not suggesting some wholesale movement back into the closet. But any movement working on behalf of the "queer community" writ large is either grappling with the question of how to help those who cannot or will not come out, or abandoning some of our most vulnerable members.

Selling Progress in Appalachia

Ruy Teixeira

A MAN WEARING a canvas shirt tucked into blue jeans walks through the tall grass, clicks a break-action shotgun, and fires a round. "I approve this message," he says. The camera zooms in on the target as it explodes in the distance.

It is not a TV ad for a Republican, though a similar shotgun did appear in a controversial ad run last fall by Georgia's recently inaugurated Republican governor Brian Kemp. The scene comes, instead, from a spot for the senior Democratic senator from West Virginia, Joe Manchin, who was reelected in November. The target he shoots displays a sign saying "lawsuit on coverage of pre-existing conditions," referring to a 2018 case brought against the Affordable Care Act (ACA) by two governors and eighteen state attorneys general, among them Manchin's Republican opponent, Patrick Morrisey. Before Manchin fires the round, he calls Morrisey "dead wrong." In this genius—and electorally effective—feat of political messaging, Manchin packages a liberal policy idea in a conservative cultural form.

The lesson is an urgent one. Elizabeth Catte suggests that the left can do better in Appalachia by appealing directly to class interests, rather than simply trying to be "moderate." I don't doubt this is true, as it is in many parts of the United States today. The question is *how* to sell this message in specific political and cultural contexts, including those in Appalachia.

THE ANSWER IS perhaps not as simple as Catte makes it out to be. She takes issue with the suspicion of Nancy Pelosi and Steve Israel that "the emerging left platform . . . might perform well in urban centers but not so much elsewhere." But Pelosi and Israel make a fair point—backed up by voting and public opinion data—about political messaging. The first rule of marketing is that you play to your audience.

Take "Abolish ICE," for example. It is difficult to see how such a demand, supported by only about a quarter of the U.S. population, could possibly gain traction in a relatively conservative—and not particularly immigrant-friendly—part of the country.

Other parts of the platform Catte discusses are probably more viable in these areas. Medicare for All, for instance—whether it is called that or something else—would likely have more purchase. There is clearly an appetite across the country for substantially expanding the ACA, not getting rid of it. There is good reason to believe that this appetite extends even to conservative, economically hard-pressed regions such as Appalachia. The same could be said about robust jobs programs and free or greatly enhanced access to some colleges, proposals that should probably be tied to others—local infrastructure, community services—and promoted as a large-scale development program for areas left behind by economic growth. More generally, Democrats should seek to advance a strong economic populist program in an area such as Appalachia.

But Democrats do need to be careful about which programs are advanced and how they are packaged. Republican tax-and-spend attacks still have considerable force, and it will not be easy to overcome them. Importing the left's maximum program into these areas is likely doomed to failure; one must pick and choose one's battles. The left just cannot afford to ignore strong conservatism on social issues, frequently twinned as it is with suspicion of immigrants and nonwhites. Appalachian Democrats, in particular, cannot run on social positions as liberal as those of their confreres in blue metropolitan areas.

And it is not just that certain messages must be moderated. There is also the all-important question of the messenger. As Andrew Levison argues in *The White Working Class Today* (2013), one of the most fundamental problems Democrats have with the white working class is that at the national and state levels, with some honorable exceptions, they typically run candidates who are out of touch with these communities (almost never hailing from them) and, worse, who project disdain for them (wittingly or not). Voters return that attitude in spades, anathematizing Democratic elites as self-serving politicians. Fixing this problem will require a course correction from years of institutional decay for Democrats at the local level and a concomitant de-prioritization of these voters by Democratic leaders. The first step to reaching many of these voters is convincing them that even if you have never been part of their communities, you understand them and do not look down on them.

Manchin offers one model for how to do that. Catte takes him to task, but his moderation on social issues may well be necessary. Manchin may have some latitude for moving to the left economically (though energy issues will pose a sticking point), but West Virginia's social conservatism—about 60 percent of the electorate there supported confirming Brett Kavanaugh to the Supreme Court, far above national figures—does mean that Manchin needs to tread lightly on noneconomic issues.

Recently reelected senator Sherrod Brown of Ohio provides another instructive example of how successful Democrats in conservative areas adapt to the political realities of their states. Brown has bucked the tide in Ohio by remaining popular even as other Democratic politicians there have failed. He has prospered as a robust economic populist, including on trade issues, and his down-to-earth persona helps win him a hearing among white working-class voters, who give other Democrats the cold shoulder. (Consider, by contrast, the unsuccessful campaign for governor by Democrat Richard Cordray, who projected a Washington-centered nerdiness and was most comfortable reaching out to voters in Ohio's big metro areas.) Brown has also been cautious about signing onto Medicare for All, though he is on record supporting Medicare expansion for those aged fifty-five to sixty-four. And while his positions on same-sex marriage and other social issues are fairly liberal, he is careful not to put these issues front and center or portray them as the cornerstone of his Democratic identity.

Still, I would agree with Catte that Manchin, in particular, probably belongs to the Democratic past in these areas. The future likely lies with a new breed of candidates who are more forthright on a populist, progressive economic agenda and somewhat less cautious on social issues. In that sense, Sherrod Brown and his artful economic populism provide a better model for the Democratic future in Appalachia.

IT IS FAIR to say the Democratic Party as a whole is moving to the left, propelled by a sweeping indictment of economic inequality and united on core social issues such as opposing racism, defending immigrants, promoting LGBTQ rights and gender equality, and advancing criminal justice reform. Democrats today—aspirants for the 2020 presidential nomination, especially—are far more willing than Democrats of even ten years ago to entertain and endorse big ideas. Taxing the rich is in; worrying about the deficit is out. The center of gravity of the Democratic Party has decisively shifted from trying to assure voters of fiscal and social moderation, as it did under Bill Clinton and to some extent even under Barack Obama, to promising active government in a wide range of areas.

These developments will have, and should have, an analogue in Appalachia. But there should be differences in how this shift to the left manifests itself in New York compared to West Virginia. Catte and likeminded activists forget this at their peril.

What We Talk About When We Talk About the Working Class

Elizabeth Catte

IN DISCUSSING THE POTENTIAL of the left in rural spaces, we have un-covered an equally thorny topic: What, and who, do we mean when we talk about the working class? We have, rightly in my mind, assessed the left as a coalition of people and politics that aims to build working-class power, but this volume also speaks to a wider conversation about the nature of that power and how it will manifest.

Nancy Isenberg's response shows us the danger of creating simplistic avatars of the working class, and she finds fault among both progressives and conservatives. The most damning exhibit she produces is the celebrity of J. D. Vance. But she is also right to pause and examine the ways the "star power" of left political figures such as Alexandria Ocasio-Cortez crowd our field of vision and absorb gendered expectations that are difficult to reconcile. "Mother Jones wouldn't stand a chance of getting elected today," Isenberg writes, underscoring the chauvinism that continues to animate segments of Appalachia's working class.

Jessica Wilkerson's and Ash-Lee Woodard Henderson's essays hint at the problematic way we conceptualize the working class in Appalachia.

Coal miners have become the region's most scrutinized political actors and have, for too long, obscured the contributions and realities of women and people of color.

The tendency to equate labor conflict with blue-collar, male-dominated industries such as mining is perhaps why Michael Kazin finds no evidence that "a mass of Appalachians are busy creating a new movement to emulate the labor-centered one that disappeared decades ago." I think more than 20,000 striking teachers fit that description. Kazin is right that we find ourselves in a much different time and place from the heyday of the United Mine Workers and their vision of organized labor. But he may overstate the degree of the rupture: although coal is dying, its extractive logic persists in natural gas, a fuel that West Virginia delivered a record 1.6 trillion cubic feet of in 2017. States continue to wrap their economies around the satisfaction of tax-fearing power players. This extraction 2.0 era has led to a very different kind of labor conflict since state austerity measures disproportionately impact the public sector and those who work there, namely white women and people of color. Instead of industrial workers feeling the brunt of mismanagement, it is public sector employees—including the agitating teachers in Oklahoma and North Dakota—who struggle.

Elaine C. Kamarck's response builds on the problems inherent in this murky transformation. In the era of "plundering bosses," the working class had a much more direct path to power. Today this path is increasingly shaped by shifting political alliances, tax incentives, and schemes that range from focused, such as the left's Green New Deal, to incredibly vague, such as bipartisan efforts to assimilate former blue-collar workers into the technology sector. Against whom do we organize if not bosses? Kamarck suggests politicians, but she also acknowledges that it is difficult to produce an example of "paradigm-shifting" political work among our most powerful representatives.

This dilemma is taken up most forcefully by Matt Stoller, who charts Democrats' failures to build institutions and execute policies in the interest of the working class. I disagree with Stoller that far-left activists are confusing the woes of temporarily unfiltered markets with the problems inherent to capitalism in all its forms, but I grant him the point that discussions of dissent must weigh tangible evidence of change.

So how do we define change? The election of better representatives, as Kamarck and Stoller suggest? The creation of more equitable institutions, as Kazin demands? Or the more honest and sophisticated assessments of class in our political language that Isenberg asks for? Ruy Teixeira finds change inevitable, but assesses its arrival at a slow and uneven pace; white, rural voters might tolerate Medicare for All, but will surely grow cold at the idea of confronting U.S. Immigration and Customs Enforcement (ICE). But Henderson reminds us that rural spaces have much to teach in that regard. In 2017 ICE conducted its largest-to-date raid on a community in Morristown, Tennessee. Residents responded with such swiftness and purpose that the *New York Times* labeled it "a town that fought back." Henderson's essay shows us how long histories of community organizing enabled that response, moving us beyond a discussion of change toward liberation. Hugh Ryan's response also presents a rebuke to the idea that, in rural places, "change comes slowly or not at all." Ryan's evidence is the proliferation of queer communes in rural spaces, along with art and documentary projects that celebrate queer rural life in all its forms. These projects do not necessarily aim to effect mainstream change, such as expanded legal protections for LGBTQ individuals, but instead offer utopian visions for liberation.

Bob Moser forcefully strikes through the question of "tangible" change and bends our lens toward radical transformation, instructively citing regional political actors such as Chokwe Antar Lumumba in Jackson, Mississippi.

Moser also points to the election of Justin Fairfax, who took the oath to become Virginia's second African American lieutenant governor holding an ancestor's freedom papers. But taking that oath meant that Fairfax would work for an administration that plans to sacrifice a community founded by freedmen (Union Hill) to an energy company (Dominion Energy) for the construction of a natural gas compressor station. To his credit, Fairfax broke rank on this issue to the extent that his position allowed, but it is this kind of dilemma—plantation politics, meet eminent domain—that moved me to stipulate that an old war continues in the present.

The peril that Moser indicates, however, is not so much acknowledging the existence of this war as bordering it with nostalgia. Nostalgia is something of a verboten emotion to critical thinkers—cloying and unhelpful—but I organize with people who live in disappearing worlds and are processing painful separations every day. Their livelihoods, yes, but also familiar landscapes that have been ruined by extraction, and, most grimly, the thousands lost to the opioid crisis. I surrender to nostalgia as an operating force in their lives and try to bend it to what Svetlana Boym might call "reflective nostalgia," which "reveals that longing and critical thinking are not opposed to one another, as affective memories do not absolve one from compassion, judgment, or critical reflection."

What people desire from the past often reveals what they hope for the future. Do people seek histories of dissent because they are sentimental and tragic, as Moser suggests, or because it helps them bridge the space between what was, is, and might be? The power of the poor and working class is in a moment of twilight. Is it romantic to observe that the sun rises after it sets, or is it realistic? The answer is in the eye of the beholder, but it seems we are all in agreement that we must be prepared to work in the dark.

Essays

Teachers with Guns

Thomas Baxter

Some names and identifying characteristics have been changed.

"I'm a teacher," I mumble under my breath. The instructor yells another command, and we collectively pull our triggers, setting off an angry crackle of handgun fire. Twenty-three paper intruders recoil quicker than senses can register. The entire scene has the atmosphere of sport; the targets do not bleed or shoot back. Squinting through the sun's glare, I look for the impact point, the void that would bleed the life from my hypothetical foe.

"This person is killing your students!" an instructor berates, fuming at our inadequacy.

The humanoid targets are faceless, sexless, standing over six feet tall. An hour before, the instructors informed us that most school shooters are male students. But few students, even high school males, are this tall. On the range this comparison is unspeakable, but I can't shake the thought: we are being trained for the contingency that we have to kill a student.

"Fire!" the instructor yells again. The barrage continues.

Standing on each side are my colleagues in public education. Teachers and administrators all—but this week, we are recruits training to prevent a school shooting. We are learning to use a gun and, if necessary, to kill. This last part is never spoken, betraying our instructors' fear that educators may not have the meddle to take a life. On this first day of training, feeling utterly out of place, I am apt to agree.

A FEW MONTHS EARLIER, my district decided to arm staff members. According to a novel reading of Ohio law advanced by the state's attorney general, Mike DeWine, in 2013, Ohio school districts have always had the option to arm teachers and do not need to make that choice public. Therefore, it is difficult to know how many districts have in recent years availed themselves of this option. Buckeye Firearms Association, an Ohio Second Amendment lobbying organization and PAC, claims that sixty-three of the state's eighty-eight districts now have armed staff.

My district's school board struggled with the decision. After the Newtown shooting, parents came to board meetings demanding to know how the district would protect their children; locked doors and security cameras no longer allayed their fears. It was easy to see that in recent school shootings, similar safety measures had proven ineffective.

Calls were made to neighboring districts and eventually Buckeye Firearms Association was asked to make a presentation to the school board. The presentation sharply divided the board's members. For some, the choice was easy; for others, agonizing.

"How do you know that this person we arm won't go off?" one board member shouted at no one in particular.

"How else can we protect students on our budget?" came a response.

I sat in the meeting, unsure how I would answer either question.

In his navy suit, Buckeye's expert emphasized that it was a district decision and that we needed to do what was best for our students—even as his tone betrayed his conviction that every school in Ohio should have armed staff. He offered polished, if often unrelated, responses to the board's questions: "It takes rural police an average of fifteen to twenty-two minutes to respond," "Most school shootings last less than five minutes," "Newtown was over in three."

Before leaving, he placed a page of statistics before each member. The board gazed at the sheet in silence as he departed, then voted unanimously to send a staff member to the summer training.

Administrators were approached first. We were given the opportunity to opt out, to choose the sidelines. One principal declined, but I felt I had to go, not only for the students, and not only because it was expected of me, but because I felt that I needed to evaluate the training for myself from the perspective of an educator. More than that, I thought: better me than someone else. I felt safer doing it myself than letting the job fall to someone I would have limited say in choosing.

I signed a confidentiality agreement. The hope was that secrecy about who was carrying would prevent me from being targeted during a shooting. I received a list of the gear required for training and was reimbursed by the school for the purchases.

The white paper from Buckeye recommended Glock handguns. Glocks are almost guaranteed never to jam or misfire, it claimed. Having never owned a semi-automatic handgun, I simply took that advice. The district supplied a thousand brass-cased 9mm rounds for my training. After that it would provide hollow points, rounds designed to swell upon impact to ensure maximum harm.

THE TRAINING is at a police firing range in northern Ohio. Rittman is the sort of town we all traveled from: rural and with little diversity. A

Rust Belt burg with sulfur and coal dust in the air, few businesses remain in the small downtown. The locals are pleasant and pleased with our arrival. We are their stimulus plan, arriving in waves every two weeks.

The range is covered in pea gravel and shell casings, and is encircled with earthwork embankments that stand eight feet high and are thick enough to stop projectiles. It is isolating, like being at the bottom of the sea. Rising from the graveled seafloor is a pole barn which serves as a classroom. At this facility, we learn a new choreography from police veterans who spit their fevered instructions. The training is effective, and our progress is quick.

After the day's shooting practice, we go inside the barn where Chris Cerino, the program's creator and lead instructor, shows us videos. Chris is a hero to local law enforcement and is held in immediate and pious reverence by the new trainees. Chris is a world-renowned competitive marksman and two-time runner-up on the History Channel's *Top Shot*. He is the closest thing in this area to a reality television star.

Red and blue lights from Chris's videos fill the darkened barn as the distorted sound from body- and dash-cam videos blasts over small speakers. Each clip ends with shots fired and a policeman killed or wounded. Between clips, Chris explains the officers' mistakes. The recurrent theme: "Don't provide too many warnings. Determine the threat, then shoot or be shot."

A fellow recruit, Anthony, is brought forward and handed an airsoft pistol. He appears gangly and awkward next to Chris. His shoulders slump forward slightly and his shirt hangs loose while Chris stands rigidly at attention. Chris is right off of a recruiting poster; his barrel chest and square jaw give him the appearance of a comic book character. He makes Anthony look like Ichabod Crane.

Chris tells Anthony to raise the gun and fire at him. They stand, arms at their sides in an uncomfortable silence while Chris beckons

Anthony to shoot him. Finally, Anthony raises the gun and shoots. Chris tries to anticipate Anthony's movements and shoot first in what resembles a Wild West gunfight. After multiple attempts, it is obvious that even Chris can only achieve a tie, at best.

Chris says, "Reaction is slower than action, every time." The adage will be repeated daily as a reminder to act before you are acted upon.

Chris starts up the videos again. They often end with officers screaming into their radio, sometimes for help, sometimes with last words for loved ones. The tone in the classroom changes after the videos, the playful banter extinguished. The day no longer resembles a professional conference, but a discourse on survival.

Chris's assistant goes over statistics from recent shootings and then praises how the curriculum will work upon us. "From this moment forward," he tells us, "you will canvas for the uncommon. You will view strange scenes as an indication of possible violence. You will never be afraid to act."

The language is militaristic and foreign; a sense of dissociation washes over me, a kind of shock. As I sit listening to the assistant, my life as an educator seems distant.

WE SPEND most of the second day shooting the paper villains. During breaks the instructors replace the tattered targets and we begin again. Blisters form on my hands early in the day, making each percussive recoil a test in pain management. I try to hide the wince that follows each shot; the atmosphere is full of male bravado, despite the participation of a number of female recruits, and visible weakness feels inexcusable.

Adding to the surreal quality of the day, a CBS news crew follows us around. The instructors say NPR will be in later in the week. When we are not shooting, the CBS crew asks questions while the program leaders troll for the next person to interview. When it is my turn, I feel the need to defend the program and my participation in it. It makes my interview stilted, which adds to my already heightened anxiety.

That afternoon, my accuracy becomes a problem. I am exhausted and my hands raw, tinted with dried blood. I see Chris walk toward me, scuffing the gravel beneath his boots until the crackling slows, then stops over my right shoulder.

"Let me see your gun," he finally says after I pull another shot. "Keep looking at the target," he says, taking the weapon.

I think about his stunt the day before, when he drew his weapon and shot upside down, pulling the trigger with his pinky. The demonstration was meant, I assumed, to show a recruit how poorly he was shooting. The result was demeaning more than anything.

I face the target until he returns the gun to me. I start firing again. Unsure of what he wants, my accuracy suffers further. When I squeeze off the third round, nothing happens. Nothing, that is, except for my body jolting forward in anticipation, and the "click" of the gun. Chris has replaced the third bullet in the magazine with a fired casing and I, in turn, have committed a cardinal sin. Jumping in anticipation is a sign of timidness, a mark against your manhood.

"No, no, no," Chris yells, becoming so animated that recruits on each side of me stop and look. "You are anticipating your shots! No wonder you can't hit shit!"

"Sorry," I respond. "I don't know why I started that."

Chris stops his wild gesticulating, shaking his head slightly before allowing a grin to twist his mouth.

"I do," he says. "It's an explosion right next to your head!" He lightly pokes my temple with his index finger. "This shit ain't natural!" With that, he pats my shoulder and heads off.

Baxter

The guy next to me smiles and shrugs. I watch Chris walk away, surveying the firing line, and think, "You're right, this shit isn't natural."

WE CONTINUE TRAINING for three more days, eight to twelve hours a day. We will be taking the Ohio Peace Officer Training Assessment at the end of the course. It is the same test required for Ohio police officers, but they are required to hit twenty-six of twenty-nine targets. As teachers, in a school setting and on the cutting edge of political upheaval, we are required to hit twenty-seven.

The group is visibly nervous upon hearing this, but by the end of the week we feel more secure in our abilities. The constant drilling has improved our skills; we begin to resemble the cops who train us. Our confidence grows.

Early in the week, the instructors seemed to yell constantly. They were regularly on someone for not following directions or not getting the procedures down as quickly as the rest of us. By Friday, though, they lighten up and share the occasional joke. This feels like an achievement. More significantly, we begin to feel as if we could defend our students.

THE VIDEOS, instruction, and repetition play a trick on my mind, though. I start to think in terms of students and attackers, those I would protect and those I would kill. The latter are strangers— unnamed, faceless adversaries like the targets. My daydreams are no longer of classroom visits, sporting events, and kids making out in the halls. They are all adventure stories, and I am always the hero. An attacker is never one of my students. I never have to shoot one of my students.

The training encourages this result. Everything about its vocabulary is designed to dehumanize our aim. The instructors' military language—"soft targets" and "areas of operation" for schools, "threats" for shooters, "tactical equipment" for guns—rubs off. On the final day, a pep talk analogizes students with lambs. We are the sheepdogs, charged with protecting them from the wolves.

I am aware that this is changing my way of thinking. I enjoy how I feel. It is a potent energy, a righteous virtue that seems completely earned. The training reassures me of my decision-making ability.

The other recruits are undergoing the same shift. During downtime we discuss guns: which we plan to buy next, what ammo our districts will provide us, and how that ammo impacts a body. We have become gun nuts almost overnight.

BEFORE WE TAKE the peace officer assessment, we are bused to an empty high school where we take turns playing "good guys" and "bad guys." The scenarios are designed to challenge our decision-making. Some recruits are chided for offering too many warnings, harkening back to our first day of training. Others are lectured for being too giddy with their new power. The latter are mainly people who had little experience with guns before the training; now they burst into classrooms, screaming television tropes such as "Put the fucking gun down!" In the midst of the action, I look around and think we are a group transformed, and the change feels good. I am powerful and gallant. I am a sheepdog.

TWENTY OF THE TWENTY-THREE EDUCATORS pass the assessment, and most return to schools where they carry daily. I trade in my practice rounds for hollow points, my training holster for one easily concealed beneath a suit.

My wife and I plan how we will hide the gun from our daughter. I purchase a gun safe for obvious reasons, but I also need to protect the secret. My wife, the district superintendent, and my fellow recruits from the training remain the only people who know. This secrecy, to which I had given little forethought, is uncomfortable. I regularly worry that the gun will "print" against my shirt if I lean forward or show if my shirt comes untucked.

Back at work, I walk the halls examining angles, doorways, and odd hallway configurations, just as Chris and his team instructed. When questioned by the school board, I deftly repeat the program's military terminology, and it clearly impresses. I am high on the atmosphere.

IN LATE FALL, an anonymous threat of school violence is shared on social media. It is improbable, but all threats must be handled as emergencies. The staff reacts with practiced precision, preparing students to evacuate. I begin alerting parents and working with the police.

As district staff have an emergency meeting to discuss the source of the threat, I instinctively check my Glock. It is a simple movement and not easily comprehended by onlookers: a movement to my hip, a swipe of the hands, and I am assured of its placement. Indexing, Chris called it. A term with broad meaning that includes checking if the gun is secured, loaded, available for use, and pointed in the right direction. As in a dream, I attempt to make sense of the day's strange course.

Fortunately, the attempt by the person who made the threat to hide his identity was amateurish, and police quickly find him: Jason Moore, a junior.

"Are you sure?" I reply, giving my mind a moment to catch up.

I HAD VISITED Jason's parents during his freshman year to discuss his attendance. They had moved to the district two years before. Sitting in their living room, they described their traveling business selling used furniture at flea markets. When they came off the road, they did their best to combine their campers by cutting doorways with a rotary saw. The result resembled a child's crayon outline sawed in the sheet metal, tufts of insulation peeking out.

I was saddened by the scene, but impressed by Jason's resilience. He had made mistakes in junior high—nothing terrible, just poor decisions. Now he kept to himself most days. He was not an honor roll student, but he mainly steered clear of trouble. We had worked through the attendance concerns and had found a shared love of reading. He liked Tolkien-style fantasy, and I preferred literature, but our interests intersected with Ray Bradbury and Dante's Divine Comedy. He was a respectful kid without any visible signs of aggression. He was close to making it out of the awkwardly married campers. At least, before today that was how it had seemed.

Now we sat staring at each other across the desk, the sheriff and a deputy behind him on either side.

I ask the obvious questions, and Jason casually admits to posting the threat. What's more, when asked by the sheriff, he admits to having guns at his house, though he is adamant that he would never bring them to school.

"Why, Jason?" I finally ask, unable to hide my despair.

"They didn't invite me," he mumbles, looking at his shoes.

"What? Who?"

"The others. To the party."

Jason lays out the story, how he learned belatedly that some students in his class had planned to leave early that day for a party in the woods. Jason, who had thought they were his friends, had not been told. When Jason realized he had been left out, he gave in to anger and posted a threat, hoping to ruin the other students' fun. It was irrational, as most teenage decisions are.

In minutes, our conversation is over, and I watch as he is handcuffed and stuffed into the back of a cruiser. The deputies try to hold back community members on the opposite sidewalk. Other policemen are already at the courthouse to obtain a warrant to search his family's property. The married campers, I think as the cars pull away.

I drive home in a devastated silence. I thought I knew Jason well, but I had never imagined him perpetrating a threat, or owning weapons. It was like something from TV, where newscasters narrate the steps leading up to a school shooting, how everyone had missed the signs. I imagine the shoot-out it could have been.

Riding through the dense countryside, I finally face the question that I had avoided from the beginning: was this right?

My decision to be armed in school had been made in the aftermath of yet another high-profile school shooting, and I had thought, "This is how I can keep my kids safe." The training had done its work on me, too, lifting me out of my habit of cynically questioning everything. I felt reassured that *of course, this is righteous.* But now it was no longer a theoretical question of protecting kids at any cost. The faceless target at the shooting range, so absurd in its proportions, had a face: Jason, whom I wanted so badly to help.

I TRY TO COMPOSE MYSELF before entering the house. My wife hugs me just a few steps inside, knowing from social media what has happened. I whisper, "We'll talk after I change," and head upstairs.

Standing in the bedroom, I unlock the gun safe and begin to pull the holstered weapon from my pants when my daughter yells and clumsily pops up from behind the bed.

"Did I scare you?"

I force a smile, and she climbs over the bed as I try awkwardly to slide the gun back into my pants. Like most children, she is quick to recognize deception. Her eyes lock onto my hip, freezing me in place.

"Dad," she starts, standing on my bed, "why do you carry a gun to school?"

I look down at her for a long moment, unable to find the words. I am a teacher, she is a student, how could I ever explain?

Baxter

The Most Radical City on the Planet

Makani Themba

UNTIL RECENTLY, most progressives wrote off electoral politics in the South. But before the near-wins of Stacey Abrams in Georgia and Andrew Gillum in Florida, there was Chokwe Lumumba, a radical black lawyer and organizer who was elected mayor of Jackson, Mississippi's capital, in 2013. His victory helped put Jackson on the map as a progressive city—a seeming contradiction in a state better known for its stubborn poverty, violent Confederate fan boys, and deeply entrenched black oppression.

As much shock as Lumumba's win elicited, it was decades in the making. Lumumba, a Detroit native with southern roots (like many black Detroiters), came to live in Jackson in the 1980s as part of a deliberate strategy to organize and build independent black power in the "Black Belt"—five southern states with a high percentage of African Americans (Louisiana, Mississippi, Georgia, Alabama, and South Carolina).

Black radicals had been experimenting with electoral strategies since the 1960s. In 2008 the Malcolm X Grassroots Movement (MXGM) studied the lessons learned from this work in the South and identified ways to advance movement goals. This work culminated in the 2012 publication of the Jackson-Kush Plan, which called for people's assemblies (a grassroots co-governance model), an independent black political party, and a broad-based solidarity economy. Along the way, MXGM members identified Chokwe Lumumba to run for Jackson city council in 2009. He won, and by the time he ran for mayor four years later, he was well known, with an established infrastructure to support him.

Tragically, eight months into his term, Lumumba died. When his son, Chokwe Antar Lumumba, ran an unsuccessful special election bid to replace him, many were ready to write Jackson off once more. But then the younger Lumumba, only thirty-four, returned more experienced and better organized in 2017, riding an astounding 94 percent of the vote to become the city's youngest mayor. Progressives were again pointing to Jackson as an example of how radical ideas could take hold, even in Mississippi.

The mayoral victories of both Lumumbas made international news. Most reports focused on the contrast between their progressive politics and Mississippi's long, violent history of white supremacy. Yet these wins are not the aberrations they are portrayed as being. They are rather best understood as the fruit of more than a century of radical black organizing in the South.

While his father paid homage to this radical legacy, the younger Lumumba places special emphasis on it. His aim of making Jackson the "most radical city on the planet" has garnered a great deal of attention, as has his picture-perfect First Family: he has two young daughters with his wife, Ebony, an accomplished professor and activist. Their combination of elegance, humility, and affection has led many to draw comparisons to the Obamas.

The realpolitik of governing in Mississippi is constricting in ways that progressives, especially radicals, can find difficult. What does it mean to seek justice and fair wages in a time of austerity? What does co-governance look like when folks don't share key progressive values? To what extent do you work with the master's institutions—such as public schools or the Democratic Party—because they reach the majority of our people? Operating in the belly of the beast tests us and our ideas. It also offers lessons for how to build power *and use it* to improve the lives of black citizens.

THE POLITICAL CONTEXT that sets the stage for black governance in Mississippi goes back more than two centuries. Slavery, terror, and the forced removal and genocide of indigenous people are integral parts of U.S. history, but nowhere more than in Mississippi. Cotton production grew exponentially between 1820 and 1860. It in turn grew Mississippi's wealth—and its enslaved population, which quintupled between 1830 and 1850 alone. By 1850 blacks outnumbered whites. White terrorism and violence—by law and by mob—was critical to controlling the black majority during and after legal slavery. There were more lynchings in Mississippi than in any other state, and enslaved blacks who fought against whites elsewhere in the country were often sent to Mississippi to be "broken" as punishment.

Despite this, Mississippi was the site of several black rebellions going back to at least 1729, when the French controlled it. Some of these rebellions were mounted in collaboration with indigenous peoples, most notably the Natchez nation. Although thousands of black families fled to relative safety in the North, the descendants of those who stayed are proud of their resistance. These black Mississippians will say, "We are the ones who didn't run."

Black resistance in Mississippi has taken many forms. Before he was assassinated in 1875, for example, Charles Caldwell worked as an organizer, state legislator, and captain of a military operation to defend black people in Hinds County (which includes Jackson) from white terrorists. Many of the organizing methods that we take for granted today—canvassing, targeted voter registration, marches—were developed and refined in the Black Belt. Decades before the modern civil rights movement, black Mississippians were building independent communities, protecting loved ones from white violence, and governing themselves. This legacy of resistance in Jackson—along with its large black population, its strategic position as the state capital and a major thruway, and its several black colleges and universities—helped draw MXGM organizers to the city.

Since its founding in 1821, Jackson has steadily elected Democrats, including a string of conservative Dixiecrats. It elected its first black mayor in 1997. But throughout the nineties and aughts, the elder Lumumba and his wife, Nubia—herself a savvy organizer who passed away in 2003—worked tirelessly under the auspices of MXGM to forge a strong political and cultural community that was rooted in ideals far to the left of the DNC platform. Their work targeted the city's many challenges, from high poverty and aging infrastructure to poor public health, bad schools, and a high crime rate.

What made the first Lumumba's victory such a watershed moment— for both Jackson and the nation—was that he represented more than just a black face in a high place. He represented a movement offering an alternative vision where black people—especially working-class and poor black people—mattered.

JUST DAYS after his father's memorial service, Chokwe Antar, then thirty-one, was drafted to run in the 2014 special election. It was a bruising campaign. His main Democratic opponent in the primary was Tony Yarber, a black minister with working-class roots and corporate backing who leveraged his pastoral alliances, anti-intellectualism, and age to paint Chokwe Antar as an inexperienced, anti-Christian, privileged transplant—even though he has lived in Jackson since he was a child. Yarber won by eight points, a margin delivered largely by Jackson's small but united white voting bloc. For the MXGM coalition, it was a time for reflection and rebuilding.

Yarber proved to be a lackluster mayor and was beleaguered by scandals, including a lawsuit that alleged sexual harassment. Lumumba, by contrast, served on a board to address challenges with a downtown development project, conducted legal advocacy, and provided input and testimony on a range of public issues before the city council. Residents appreciated his persistence, and he became the obvious frontrunner for the 2017 mayoral race even before announcing his intention to run.

The new campaign was a formidable operation. Though it was multiracial and intergenerational, it was primarily black millennials, especially younger black women, who served as its leaders. Ebony ran the media campaign, which included a social media effort like no Jackson campaign before it. Lumumba's sister, Rukia, served as campaign manager, placing a heavy emphasis on the ground game.

Lumumba's campaign promise to make Jackson the "most radical city on the planet" deflected the charge that he was simply following in his father's footsteps. The younger Lumumba wanted to surpass his father's vision, making a more explicit commitment to cooperatives, offering a clear plan for innovating city government and the people's assemblies (a pillar of the Jackson-Kush Plan), and committing to participatory budgeting. His campaign slogan, "When I become mayor, you become mayor" (adapted from Newark mayor Ras Baraka's), inspired Jackson residents who felt locked out of their government. At the national level, the campaign won endorsements from trade unions and progressive political organizations such as the Working Families

Party and Our Revolution. Lumumba beat eight candidates in the Democratic primary (including Yarber, who received only 5 percent of the vote) and then sailed through the final election.

At the inauguration, attendees were asked to swear a "people's oath" of participation and co-governance, agreeing to "faithfully serve as a citizen and supporter of the city of Jackson. . . . Together, we will rebuild our infrastructure. We will create jobs for Jacksonians. We will provide holistic solutions to crime, and we will improve our educational system."

LUMUMBA'S ELECTORAL VICTORY was a powerful show of unity among Jackson's black and progressive leadership. Pastors who had previously supported Yarber, for example, publicly apologized for their "error." But it has not been easy for the Lumumba administration.

Kali Akuno, a former staffer under the elder Lumumba's administration and a former leader of MXGM, has emerged as an outspoken critic. Akuno helped write the Jackson-Kush Plan and thinks Lumumba is not doing enough to engage with its main tenets—especially growing an independent black base. For critics such as Akuno, both MXGM and Lumumba prioritized electoral victories over the development of an independent black political party.

Mississippi, like most states, makes running as a third-party candidate challenging, and with the vast majority of Jackson voters registered as Democrats, the Democratic primary is now pretty much the only stage in town. Developing an independent party would require a different set of organizing and legal strategies. Of course, creative party organizing is not new to the city. Jackson was the headquarters of the 1964 Mississippi Freedom Democratic Party, led by Fannie Lou Hamer and Lawrence Guyot, which directly challenged the Democratic Party, though it was not a third party in the traditional sense.

Building "outside" formations—ones independent of mainstream parties—poses a stark challenge not just for Lumumba but for other progressive organizations: How do you win elections *and* continue to build an independent base to push the work forward and hold those you elect accountable? Few groups do both well. Many on the left who are fond of ideological purity find it hard to do the pragmatic, even messy, aspects of mass engagement: meeting people where they are, building awareness and support, dealing with limited resources, and being willing to educate across the political spectrum.

MXGM has continued to organize Jackson's people's assemblies, which have been recently revived under the leadership of the Democratic Visioning Committee (DVC). The DVC is a body of Jackson-based organizers and advocates (of which I am a part) cochaired by Lumumba's sister, Rukia, and longtime MXGM leader Akil Bakari. The assemblies, organized around small group discussion, allow residents to discuss issues, communicate priorities, and speak directly to city officials—many of whom regularly attend, including the mayor and lead staff. Much of the meeting happens in small groups to allow for deeper debate among resident. Attendees have explored participatory budgeting, which kicks off as official city policy soon, and have also taken on crime and safety. Participants tend to share progressive values and neighborly good will, but the assemblies offer a constant reminder that the city is politically diverse.

Perhaps the biggest test for the Lumumba administration has been policing—a progressive official's third rail. Jackson has the sixteenth highest murder rate in the nation per capita. Chokwe Antar ran on a platform of prevention and prison alternatives, but abolition is still a difficult topic for many Jacksonians since most tie safety to policing. A rash of police-involved shootings last spring provoked a conversation about accountability. Activists wanted more than condemnation from the mayor. They asked him to use his authority to repeal the department's policy that protected the names of the officers involved. Leadership of Jackson's predominantly black police force, on the other hand, framed the policy as a workers' rights issue.

Guided by his people's governance framework, the mayor convened a task force to review the policy, solicit public feedback, and propose changes. The process took five months—light speed for bureaucracy, but an eternity for grieving family seeking justice. In the end, the commission negotiated a compromise requiring police to release the names of officers involved in shootings within seventy-two hours, so long as the officers were not operating undercover. Jackson is the only jurisdiction in the state of Mississippi with such a policy.

LAST YEAR, the Lumumba administration presented a strategic plan framed around creating a "Pathway to Human Dignity." The document's central tenet, the "Dignity Economy," has five key components:

1) Healthy citizens
2) Affordable homes in safe neighborhoods
3) Thriving educational system
4) Growing tax base with occupational opportunities
5) A city that is open and welcoming to visitors

It is certainly not the stuff of revolutions, but it led to improvements almost immediately. Nonessential city workers were finally able to come off monthly furlough last year (they had missed one Friday each month since 2015), and in September, workers received their first pay raise in years.

Does reform have a role in radical governance? The tension is ancient. Reform is only as good as the vision and plan it serves. It is about learning how power works, creating spaces to practice making decisions together, and easing suffering by controlling what we can. It is about learning to have faith that, as people not traditionally in power, we have the answers, the know-how, and the imagination required for the challenges ahead.

Chokwe Antar threw down the political gauntlet when he declared his intent to make Jackson the most radical city on the planet. Not yet two years into his administration, few expected to see a transformation. But there have been changes. Some came immediately, including an initiative to provide living-wage jobs to formerly incarcerated residents. Some has been slower but no less visible: Jackson voters helped sweep in an entire slate of black judges in Hinds County in November. Less tangible but just as real is that people take pride in their city.

These achievements may seem like baby steps, or even detours on the road to radical governance. But anywhere that black people are working to govern together and navigate their differences for a better life for all has radical potential.

What Activists Know About Fighting the Opioid Crisis

Lesly-Marie Buer

IN 2017, for the second time in recent years, U.S. life expectancy decreased. Headlines blamed the decline on suicides and opioids, and cast impoverished rural whites as the primary victims. A great deal of attention has been focused on Appalachia, whose population is (erroneously) portrayed as uniformly white, poor, and ravaged by drug addiction. *White* sickness has thus come to stand for what is supposedly wrong with health, health care, and culture in the United States.

The truth is not so simple. Black mortality rates continue to dwarf those of whites—another tragic indication of how our society has normalized racial inequality. In West Virginia, the state with the highest overdose death rate, the rate of overdose among blacks is slightly higher than among whites. In Tennessee, whites fare worse than blacks, but maybe not for long: from 2008 to 2016, the overdose mortality rate more than quadrupled among blacks compared to about a doubling among whites. Moreover, mortality rates increased for seven of the ten leading causes of death, with the highest percentage increase seen in influenza and pneumonia. Drug addiction, then, is not the easy explanation we have made it out to be. Income inequality, loss of social safety net services, and state violence against communities of color are also massive problems, a conclusion borne out by a large body of research. But by only focusing on class—those poor white Appalachians—most media reports ignore the racism, xenophobia, and LGBTQ discrimination found in every aspect of the U.S. health care system, from medical research to bedside care.

There is hope in this story, however. Because these marginalized communities have been hung out to dry for so long, they have developed the tools that are now proving most successful in combatting the opioid overdose crisis. "Harm reduction," as this approach is called, is a social justice movement that supports the dignity and rights of people who use drugs. Many harm reduction programs have roots in 1980s activism around HIV, and most of the organizations I have worked with exist thanks to the unrelenting labor of queer activists. Some view the approach as controversial because it bucks the dominant neoliberal attitude toward drug use, exemplified by the War on Drugs's focus on punishment. Instead of trying to enforce abstinence—viewing individuals as the site of punitive state intervention—these initiatives try to reduce the harms, both medical and social, associated with drug use.

With social services and medical support increasingly out of reach for many Americans, harm reduction takes seriously the intersecting inequalities that push people into the fringes and drive down life expectancy. If efforts to combat overdose deaths are to be successful and equitable, we must acknowledge how these inequalities affect people who use drugs—within and outside of Appalachia.

Drug use was a part of my life long before it was a research interest. I learned about Oxy, meth, and all the rest from hanging out in people's houses and cars. Many of these people have tried abstinence-only programs repeatedly. Every unsuccessful round brings not only another layer of shame but also an increased chance of overdosing, since any period of abstinence diminishes one's tolerance. When I finally began researching opioid and meth use as an anthropologist, I came at the issue with a simple desire: to help the people around me live.

Three main service provisions fall under the harm reduction umbrella. First, ensuring access to medication-assisted treatment (MAT) such as methadone or buprenorphine (commonly called bupe). These drugs are opioids, thus fulfilling the body's biochemical addiction, but they are carefully regulated and using them is associated with better outcomes than abstinence programs. Not everyone does well with MAT, but those who do thrive. As one woman told me, bupe "keeps me a functioning human being. I'll stay on it the rest of my life if I have to."

The second service provision is easy access to the drug naloxone, which counteracts an opioid overdose. In 2017 there were over 70,000 overdose deaths in the United States, the vast majority of which involved opioids. Although documented naloxone saves are likely underreported due to a variety of issues—including a lack of a centralized reporting system and an unwillingness of some people to acknowledge saves—we know at least 26,000 lives were saved by administering naloxone between 1996 and 2014.

The third pillar of harm reduction is "safer consumption." This includes syringe services programs (SSPs), where used needles can be exchanged for sterile ones—thus decreasing the odds of contracting HIV, hepatitis C, and other infections—as well as the more controversial safer consumption spaces (SCSs), often called supervised injection sites. Many people without access to SSPs develop abscesses because they reuse needles and are wary of being stigmatized by health care providers if they seek help. With SSPs, they are not only able to obtain clean equipment but also have a contact point to ask about wound care, to assess when clinical care is necessary, and to connect to nonjudgmental services.

All three of these strategies are associated with better outcomes than abstinence-only programs. Yet they have been slow to catch on due to a combination of cultural preconceptions and access challenges, especially in rural areas with high poverty rates. Moreover, the relationship between harm reduction and the state is tricky. While government service provision and funding can obviously help, they also heighten surveillance, which can further marginalize people. Participants in state-funded substance use treatment programs sometimes describe them as "setting people up to fail" because the programs are incredibly strict and often do not have the resources to provide quality care or to ameliorate underlying socioeconomic inequalities.

For all these reasons, many harm reduction strategies are still heavily regulated, if not outright illegal. For those of us working on the ground, the threat of incarceration or program closure looms over our attempts to save lives.

IN RURAL KENTUCKY and Tennessee, the most widespread harm reduction strategy utilized thus far has been naloxone distribution. Prescribers, local nonprofits, and municipalities dispense doses to first responders as well as lay people. It is not an over-the-counter medication, but all states have taken measures to dispense naloxone without a patient-specific prescription, meaning it is at the pharmacist's discretion. And it is still very much a *choice* to distribute. Many counties do not have a single entity that is willing to distribute naloxone. When asked if she distributed naloxone, the lone pharmacist in one East Tennessee county replied, "We don't do that here."

The cost of naloxone is also often prohibitive. The most common form is Narcan, a nasal spray, which costs around $150 for two doses without insurance—and those most in need of the drug are often uninsured. Moreover, more than one dose may be required to counteract an overdose, and if fentanyl is involved, upward of three or four doses may be needed. While cheaper generic naloxone is available, it is injectable and more difficult to administer in an emergency situation.

MATs face similar hurdles in rural areas: demand far exceeds supply. Methadone has been available in urban settings for decades but still has little reach in rural communities. Federal and state requirements stipulate that methadone participants consume their daily dose in front of licensed medical staff, and there simply are not clinics with sufficient staffing in rural areas. Bupe, on the other hand, is a more easily distributed MAT; any physician who takes a short Drug Enforcement Agency course can prescribe take-home doses. Access, however, is still limited because physicians can only prescribe bupe to 30 patients in their first year of prescribing. If they meet certain requirements, a prescriber's limits can then be increased to 100 and then to 275 patients. But, again, health insurance coverage is a barrier, which is why the case studies of Kentucky and Tennessee—which have taken different approaches to Medicaid expansion—are so illustrative.

Kentucky expanded Medicaid under the Affordable Care Act, making substance use treatments easier to afford. The state's uninsured rate for opioid-related hospitalizations dropped 90 percent from 2013 (the year before expansion) to 2015 (the year after implementation). In states without Medicaid expansion, the rate dropped only 5 percent. During this same time period there was a 700 percent increase in the number of Kentucky Medicaid beneficiaries utilizing substance use treatment. Nationwide, Medicaid expansion is associated with increased bupe prescriptions without a corresponding increase in opioid pain reliever prescriptions—a sign that people are seeking treatment instead of fraudulent prescriptions.

Kentucky city mayors, judges, and other local politicians are increasingly coming out in favor of these programs, making them sound more like FDR Democrats than the staunch conservatives they had claimed to be when it came to health care. They cheer on Medicaid expansion despite the Republican governor's tirades against it, and they support bupe and SSPs, often commenting that they know someone affected by the new programs. They also endorse reforming the criminal processing system, because they see how county jails have become costly holding facilities for people with substance use and mental health concerns. The jails often offer no treatment and minimal health care, leaving people to withdraw, and sometimes die, in their cells.

Kentucky had a Democratic governor to help usher in Medicaid expansion in 2014, but next door, Tennessee's Republican-controlled state legislature has consistently blocked expansion efforts. While drug overdose deaths have risen nationally, they have risen more sharply in states without expansion. There are four more overdose deaths per 100,000 people in non-expansion states. Medicaid spends dramatically more on naloxone in expansion states—a sign both that it is being used and likely saving lives. Yet some Tennessee conservatives would rather keep their heads in the sand. When the Republican governor proposed a Medicaid expansion in 2015, the Beacon Center of Tennessee, a right-wing think tank, took credit for defeating the proposal in committee. Last year Tennessee Republicans doubled down, joining three other states in pushing for Medicaid work requirements, which would almost certainly purge

TennCare Medicaid roles of thousands of recipients. At this point in the opioid crisis, however, it is difficult to see how long a strategy such as this can last. Many mayors, city councils, and even district attorneys have seen the need for more harm reduction strategies in the state. In the words of one county commissioner, "How can the opioid crisis be part of their campaign if they're not going to support a syringe exchange?"

Tennessee does, in fact, have a syringe exchange program—it was legalized in 2017—but most residents know little about it, and few have used it. The state's SSP legislation is burdensome: only state-approved nonprofits are allowed to administer the programs, which greatly limits the availability of such services in rural areas. In the almost two years since legalization, only four programs have been launched in the state, all in cities. Residents of East Tennessee's rural counties must travel one to two hours if they want to access the closest SSPs. (Prior to SSP legalization, a local "friendly pharmacy" in one county took matters into its own hands, collecting and disposing of used equipment while providing unused syringes.) In Kentucky, by contrast, where a 2015 law legalized SSPs and allowed county health departments to operate programs, there are now forty-five programs spread across the state, serving rural as well as metro counties.

Despite these successes, Kentucky also shows how the government can inflict harm rather than alleviate it. When bupe clinics flourished in 2014, local law enforcement and Child Protective Services (CPS) read the demand for it not as an unmet medical need, but rather as recreational abuse—that people who used drugs were just trading one drug for another. Value judgments in this line of work are never far from the surface; I have heard state officials at all levels spew vitriol about bupe. One court administrator in an East Kentucky county spread rumors (untrue, as far as my research revealed) that a local clinic was illegally writing lifelong bupe prescriptions without a clinician. That bupe program closed several months later, leaving the county without a clinic.

While Kentucky state agencies created policies to fund and condone bupe, local agency actions often undermine the efforts. Because providers fear that local police will harass them and their clients for any small misstep, they run their clinics along strict guidelines. People are quickly kicked out of programs if they do not meet counseling requirements or if they test positive for other drugs—even substances such as cannabis that do not interact fatally with bupe. These dismissals can be catastrophic for clients who are left to withdraw from bupe on their own. Given the pain of withdrawal, many, of course, turn to illicit markets, whether for bupe or other opioids.

Even those who maintain prescriptions may still be targeted for state abuse. One CPS administrator told me that she hated bupe and had begun automatically removing children from their mother's custody if the mother tested positive for bupe during pregnancy or delivery, even if the mother had a prescription. Her actions ran contrary to her agency's own procedures and may have been illegal. But she told me she thought she was doing the right thing and that regional and state administrators would not notice—or interfere with—her actions. Such vigilantism flies in the face of public health research. For those mothers with active CPS cases, medication-assisted treatments such as bupe are associated with better health outcomes as well as higher rates of family unification. But many mothers I have talked to are wary of taking it because they fear the punitive hand of CPS and law enforcement. Instead, they accept the increased risk of relapse and overdose.

Mothers who use drugs are some of the most vulnerable people in the United States, targets of both public and private oversight. North

Carolina's Project Prevention, for example, pays primarily black, low-income women who use drugs to become sterilized. Tennessee's fetal assault law, which expired July 1, 2016, prosecuted and jailed women if they were pregnant and tested positive for drugs. These retributive actions of the neoliberal state stand in stark contrast to the aims of harm reduction and reproductive justice.

WHAT IS TO BE DONE? I have watched this crisis unfold both from within a state department of health and alongside radical activists. Medicaid expansion and SSP legalization are the lowest-hanging fruit when it comes to combatting opioid overdose deaths, but we must also keep our eyes on the more difficult work of combatting rampant inequality. Despite the efforts of queer activists, most substance use treatment options in rural places such as Appalachia belong to a culture pervaded by evangelical Christianity and hostile to LGBTQ populations. And despite similar, if not lower, rates of drug use compared to whites, black and Latinx populations face much higher incarceration rates for drug-related crimes.

As we continue to fight these systemic injustices, we may look to a tradition of underground services for inspiration. The state has an important role to play, but learning to care for oneself and others using limited resources—especially in rural areas—is equally important. From those "friendly" pharmacies that provide syringe services to "street docs" armed with naloxone, people will provide the care they can to save themselves and others, no matter the law. The existence of safe consumption sites across the country is a case in point. These sites have operated for decades. Some are highly organized, relying on dedicated volunteers, including nurses and doctors, to provide SCS services at regular times and places. Others are more informal. One person I spoke with described a group who met in an abandoned trailer every evening in eastern Kentucky. The meetings are not only social, but protective. Word has spread: these efforts save lives. Today, cities such as Philadelphia, San Francisco, and Denver are attempting to open formal, city-sanctioned SCSs, despite the Justice Department's vow to shut them down.

In the absence of more government resources or funding, those of us on the ground need to continue our work, all while pushing to decriminalize our actions. Public health workers and activists often see politicians and law enforcement flinch when we mention these efforts, but we cannot allow this to discourage us. The simple truth is that harm reduction is evidence-based public health practice. It is time the country takes it seriously.

Every Crucifixion Needs a Witness

William J. Barber II

interviewed by Toussaint Losier

WHEN I TOLD the Rev. Dr. William J. Barber II that I also had family from small-town eastern North Carolina, he was delighted. "My people are from there," he chuckled. "We might be cousins!"

This sense of the interconnectedness of rural life in the South, reflected both in personal genealogies and political histories, has served as the backbone of Barber's call to rebuild the struggle for social justice on a moral foundation. The longtime pastor of Goldsboro's Greenleaf Christian Church, Barber was president of North Carolina's NAACP for more than ten years. During this time he helped to launch the Forward Together Moral Movement, which gained national attention for its Moral Mondays protests at the North Carolina General Assembly. Amidst the hard-right, Tea Party–style takeover of state government, this movement used civil disobedience and coalition building to combat a variety of injustices such as voter suppression, environmental devastation, and cuts to social welfare programs.

Last year Barber joined with others in reviving Martin Luther King, Jr.'s Poor People's Campaign in an effort to address poverty as a moral issue. In October 2018, the MacArthur Foundation recognized his commitment to building progressive movements and "broad-based fusion coalitions" with its prestigious Genius grant.

Grounded in a deep appreciation for movement history, Barber's goals and methods upend conventional political axioms, and his successes make a compelling case, as he says, that "the solid, red South is very vulnerable"—a prospect we discussed in the following exchange in mid-December.

TOUSSAINT LOSIER: In your book *The Third Reconstruction: How a Moral Movement Is Overcoming the Politics of Division and Fear*, published the year Donald Trump was elected, you detail how years of pastoral care and community organizing in rural Virginia and North Carolina led you to discover that "fusion coalitions rooted in moral dissent have power to transform our world from the grassroots community up." Could you explain what you mean by "fusion coalitions"?

WILLIAM BARBER: When we look at history, we find that almost every progressive achievement—from abolition to social security to civil rights—grew out of some form of fusion politics with a moral under-pinning, not a policy underpinning. When white abolitionists and black abolitionists, slaves and former slaves, Quakers and white evangelicals of that day (not the kind we talk about now) fused together, they were able to create an abolition movement. When King talked about the Poor People's Campaign, he was talking about fusion. He said it was his campaign, but actually he pooled together twenty-five groups—from Jewish groups to welfare rights workers—in the initial meeting.

So, for me, fusion is when people are brought together, and they fuse their ideas, their intellect, their power, but they do it from a moral critique. They look at society through the lens of our deepest moral issues. In 2008 my organization mapped the way North Carolina politicians voted in the state legislature. We found that the same people voting against environmental justice were also voting against public education. And the same people voting against public education were also voting for voter suppression laws. We realized that these interlocking injustices required an intersectional response. We were able to show that poor black people and poor Latinx communities and poor white communities and poor Native American communities were all being hounded, haunted, and hurt by the same politicians.

And this was when Democrats were in control! I want to emphasize that we didn't start the Forward Together Movement when the Republicans were in office. For years we had been fighting Democrats to get same-day registration and early voting passed in North Carolina. For years, when we worked in our silos, it didn't get done. Once we came together with all our partners and tacked our agenda on the door of the legislature like Marin Luther with the Ninety-five Theses, that's when we were able to get same-day registration and early voting passed.

TL: The Moral Monday protests, where you and others enter the North Carolina legislature peacefully and are then arrested each week, started in 2013. What has it meant to have that kind of ritual? That embodied practice of showing up week in and week out with the faith that you can make a difference, but relying on, as the Apostle Paul said, "the evidence of things not seen"?

WB: Well, down through history, we've never won anything with one march, one tweet, one speech—that is a misreading of history. We've always won progressive and revolutionary growth in this country through campaigns. Through 381 days of having a Montgomery bus boycott, for example, or through years of the abolition movement. Persistence in a movement is a ritual you must have. You must have a deep commitment to civil disobedience and be willing to put your bodies on the line in a nonviolent way, rooted in love and justice, to dramatize the ugliness of what's going on.

What too often happens in the United States today, particularly with some progressives, is we put too much of our actions into electoral campaigns. And when the campaign is over, and we lose, we go home until the next campaign. Even extremists don't do that. Even when they're in the minority, even when they lose a vote, they continue to organize. In some ways they've stolen from us what people such as King instructed us to do.

Now, when the Republicans took the majority in 2013, we found that they were extremists on steroids. In the first fifty days, they attempted to undo everything that we had won and pushed for since 2008. They denied Medicaid expansion to 500,000 people in North Carolina—342,000 of whom, by the way, are white. They went against immigrants, the gay community, and women. They denied a minimum wage increase, even though we had won an increase earlier. And then they went after voting itself.

So we decided then that if they were going to crucify the poor, crucify women, crucify children, crucify the sick—every crucifixion needs a witness. And they needed a consistent witness. People told us, "Well, there's nothing you can do. They have the majority. You just have to wait until the next election." But we looked at history and saw that people who made change never waited until the next election. For instance: the Voting Rights Act of 1965—there was no election that year. None of the politicians changed in Congress that year; many of them never even intended to support a voting rights act. But King, Rabbi Abraham Joshua Heschel, and the others—they changed the political atmosphere, which forced the change in the politics.

We couldn't wait for the next election. We decided with the Moral Monday Movement that we needed to have a nonviolent civil disobedience campaign every week to drive home for the public what was happening and to ensure that what the politicians did didn't happen in the dark. We believed that a lot of people didn't actually know what was going on and holding just one protest or march would not politically educate them. Doing it every Monday gave us a stage to address the issues; it created a following with the media; it created our own social media. It allowed us to shift the narrative, and let people see what was going on.

It also forced the legislature to constantly respond—and they didn't do well in their responses. After the first six weeks, Governor Pat McCroy, who signed all these regressive bills, went from a 50 percent approval rating to under 40 percent and then never recovered. The legislature dropped to 24 percent approval in the polls and also never recovered.

On the flip side, our annual event went from 20,000 people to over 80,000. We were then able to add to our efforts. Lawyers came to offer their services pro bono, and we filed a major lawsuit against voter suppression. Three years later, we won. That would have never happened if we just had one annual event. We constantly said to people that we're not building a moment, we're building a movement.

TL: And as the movement grew, you were very deliberate about drawing the distinction between a partisan movement and a moral movement. You tell the story of being invited to speak in Mitchell County, a heavily white county in western North Carolina. Can you reflect on what that invitation meant, and how you were able to resonate with folks who might initially consider themselves on the other side of fence?

wb: When we looked at these interlocking injustices, we looked at them not from the standpoint of Democrat or Republican because those are puny, small terms that keep us in our silos. Everybody uses them, but they're designed in tribalism and they're used for tribalism. People are more than that, but we still automatically look at people and pigeonhole them. Most people don't know this, but at least 50 percent of the people who got arrested in North Carolina during Moral Mondays were white—even though the movement was being led by someone who identifies as African American. (My ancestors were white and black.) But if you're black and you're talking about issues, that makes you a civil rights leader, not a moral leader. If you're pushing for voting rights, you must only care about black people voting—you don't want to talk about issues that impact all people.

What we found is that by framing our message from a moral perspective, we got people to stop and listen. Then it also opened us up to go to places such as Mitchell County that we would not have been able to go if we were just a blue movement or a left movement. So we would go to Mitchell County, which is 98 percent white and about 80 percent Republican, and say, "Did you know that your politicians are refusing to raise the living wage, and you have so many people here who need it? Do you realize that a thousand people in this county could get health care, but your elected officials are throwing up these dog whistles that would suggest lazy people (i.e., black and brown people) are going to get free stuff when, in fact, the majority of people who would benefit from Medicaid expansion in North Carolina are white?"

By having that kind of conversation, we helped people not be stuck in those puny terms, but see what was going on in a much broader way. That then allowed us to build this fusion coalition.

TL: What led you to resurrect the Poor People's Campaign nearly fifty years after King's death? What does King's vision have to offer us today, especially when it comes to the failure of our political leaders to address the crisis of poverty and the potential gains to be made by addressing it as a moral issue?

WB: Fifty years ago, King talked about poverty, racism, and militarism—the three evils. These days, we have five evils in our democracy.

One is the systemic racism that you can see through voter suppression, resegregation of public schools, a new Jim Crow prison system, and the mistreatment of immigrants and Native Americans.

Two is systemic poverty, because we don't have 40 million people in poverty, like the U.S. Census says. We have 140 million people struggling —almost 43 percent of all the people in this country—and the majority of them are white women, children, and the disabled.

Three is ecological devastation, which is paired with the fact that 39 million people right now don't have health insurance.

Four, this military war economy. We budget $700 billion every year for defense. The average CEO of a military contractor earns $19 million a year while the average private in the U.S. Army makes less than $30,000—some have to be on food stamps.

And five, we have a religious Christian nationalism—which is a form of heresy—that says the only moral issues allowed in the public square are abortion, gay people, praying in schools, and guns. Some issues are not about left versus right, but right versus wrong.

TL: What lessons have you drawn so far from the Poor People's Campaign? Where do you see it going?

WB: We've learned that we have to shift the narrative because our politics right now is not designed to address poverty and racism. And that is a fault both of Democrats and Republicans, but in different ways. Republicans often want to blame poverty and racism on personal morality, and sometimes Democrats want to engage in the kind of neoliberalism that says if you just focus on the middle class and working class, then you've fixed everything.

But almost every state that is a voter suppression state is also a high poverty state, a state that has denied health care, a state that has denied living wages. So, what we're learning and showing people is how these things are connected and why you can't just operate in your own siloes. You can't say, "Well, I care about health care, but I don't care about voting rights."

You know, when King called for the Poor People's Campaign, a lot of people walked away from it. What we're finding now is that people, including the religious community, are coming to it. There's a large number of religious bodies ready to join and engage this movement.

We're doing deep work in the South because it holds the key to the transformation of this country. If you take the southern, former slave states, that's 170 electoral votes right there. Progressives have given those states away, for the most part. But because of shifting demographics and the ability, when it's time, to bring this coalition together, the South can be changed. They're not so much red states as they are unorganized and unmotivated states.

A lot of people are just not engaged in politics. Close to 120 million people didn't vote in the midterms. A lot of them don't vote because they never hear their issues. We have to force this change of narrative. We are beginning to see this happen, even with the near-victories of Stacey Abrams in Georgia and Andrew Gillum in Florida. But I'm not talking about the individuals; I'm talking about the coalitions that came together in those states. They are showing that the solid, red South is very vulnerable—that it can be broken.

The South is not going to be transformed through electoral politics every two years. It's going to be transformed through movement building. And if we build that kind of movement—if we shift four or five southern states—we fundamentally shift the politics of the whole nation.

Barber & Losier

Bad Neighbors

Robin McDowell

ON JUNE 2, 2015, I celebrated my birthday at the Iberville Parish Court in Plaquemine, Louisiana. Plaquemine is about twenty miles southwest of Baton Rouge at a sharp turn in the Mississippi River, and I had woken up early to get there on time, filing in quietly with two friends who were there to have their case heard. We took our seats, and watched as the side door opened to admit eight men in orange jumpsuits. They were chained at the feet, hands, and necks.

"Are y'all having fun yet?" the judge asked, flashing a grin at the tiny audience.

Until his retirement in 2017, Judge James J. Best of the Eighteenth Judicial District of Louisiana presided over Iberville, Pointe Coupee, and West Baton Rouge Parishes. Locally he was most known for once grabbing a young black man by the neck during a sentencing hearing and declaring that he ought to be "taken behind a shed and whipped."

There were three cases on the docket that day, and Judge Best made quick work of the first two: he denied the probation petitions of the chained men in the front row and then dismissed a police misconduct case brought by a man who had been crippled by officers during his arrest. Then it was our turn.

My friends, Janice Dickerson and Vivian Chiphe, were up against Axiall Chemical, one of the largest vinyl manufacturing facilities in North America. The company was seeking to expand its plant onto the grounds of Revilletown Cemetery, where Dickerson's and Chiphe's forebears, some of whom had worked the land beneath the plant as slaves, were buried. Dickerson's and Chiphe's families went back to the founding of Revilletown, a small settlement created in 1874. Former slaves, organized as the Mt. Zion Baptist Association, bought a parcel of land from David Reville, a white doctor from Kentucky who owned the Reville Plantation. The present-day cemetery was established on this land, since the deceased could no longer be buried in the plantation's slave cemeteries. Revilletown's Mt. Zion Baptist Church was built a year later in 1875, separate from the Mount Zion Baptist Association.

Why does a tiny cemetery plot matter to Axiall? The company is worth billions; it is a Leviathan in no danger of shuttering if it cannot own a half-acre parcel of grass and bones. And yet here we were, three years into the legal battle and staring down a team of six corporate lawyers fighting for a company that wanted to own Revilletown's dead—the most recent of whom it may have had a hand in killing with its years of pollution. Why?

Answers begin to emerge when we consider the full history of Axiall. This history illustrates how an international network of factories, an entire town, and a patch of grass are conscripted into an economic, social, cultural, legal, and racial regime that spreads over land and across time. This regime must keep the people on whose ancestral grounds it squats exactly where they are—working for Axiall or living below the national poverty level. This regime needs to keep young black men in coffles across the river in St. Gabriel's

Elayn Hunt Correctional Center. It needs men such as Judge Best to maintain that order. It needs liberal politicians, comforting orators, and plenty of committees, task forces, and hearings to listen to and report on the needs of "the people" while turning a blind eye to the polluted southeastern Louisiana skies that send far too many to early graves. Nothing can be allowed to escape—not even a parcel of grass—because then anything could, and this wealth-generated system is built to prevent the loss of capital.

GEORGIA HARDWOOD LUMBER COMPANY, the enterprise that would become Axiall Corporation, was founded in Augusta, Georgia, in 1927. The fledgling wholesaler owned and represented five sawmills in the South. During World War II the company grew by leaps and bounds to become, according to its self-published history, the largest supplier of lumber to the U.S. Armed Forces. By 1949 the company—now with a new name, Georgia-Pacific Plywood and Lumber Company—was listed on the New York Stock Exchange with revenues of $37 million. In 1968, the year Mt. Zion Baptist Church upgraded to a new building in Revilletown, Georgia-Pacific sales hit $1 billion.

Residents of Revilletown have described their home as "poor in substance, but rich in spirit." But to Georgia-Pacific, Revilletown was, as journalist Justin Nobel summarizes, "of magnificent industrial value, as a ready-made port with quick access to the grain and corn fields of the Midwest, the oil and gas reserves of the Gulf Coast, and the deposits of salt that layer southern Louisiana and contain chlorine, a necessary component in the production of many fertilizers and plastics." And so, in the mid-1970s, a thousand feet from Revilletown, Georgia-Pacific opened its first large-scale phenol/acetone and methanol production plant. The chemicals would be used to manufacture plywood and other wood fiber and granule-based wood byproducts, notably Brawny paper towels and Coronet toilet paper. The plant provided much-needed jobs, especially as soldiers from the Vietnam War began returning home.

In 1975 Georgia-Pacific took advantage of Louisiana's underground salt domes, formed by prehistoric inland oceans. The domes were commonly used to store oil and natural gas, but the salt extracted from the nearby domes could also be mixed with water and run through electric currents to create chlorine-based compounds, caustic soda, and hydrogen. This chlorine compound operation was so successful that Georgia-Pacific added a new plant on the same site that would vertically integrate the production of VCM (vinyl chlorine monomer) with the methanol-based manufacturing of paper pulp. These vinyl compounds, particularly PVC (polyvinyl chloride), could then be used to create pipes, window frames, siding, flooring, shower curtains, bottles, medical tubing, and myriad other products.

In 1978 Georgia-Pacific added yet another plant beside Revilletown. The existing methanol and chlorine plants produced hydrogen as a byproduct. This excess hydrogen could be used to produce ammonia, a key component in, among other things, commercially produced fertilizers. By 1980 Georgia-Pacific expanded its vinyl facility and built a fourth plant, which produced sodium chlorate for bleaching paper and creating rocket fuel. Georgia Gulf Corporation, the chloride and petrochemical product manufacturing division that was mostly based around Revilletown, split from Georgia-Pacific through a management-led buyout in 1984.

About a hundred people lived in Revilletown during this time, and they began noticing high rates of cancer. Revilletown, sadly, is not unique; tragedies such as this are familiar in what has come to be known as Cancer Alley, the slice of Louisiana along the Mississippi that is dominated by petrochemical companies located in close proximity to poor, mainly black communities. In Revilletown, evidence of exposure to vinyl chloride gas was found in lab work done on the community's children, and several deadly explosions over the years prompted citizen mobilization against Georgia Gulf. The levels of toxic waste in the air, soil, and water were found to be in violation of Environmental Protection Agency (EPA) standards. In 1988 residents brought a class action lawsuit against Georgia Gulf, which was settled out of court, with Georgia Gulf buying out all of the lots that comprised Revilletown. All told, Revilletown residents received a little over a million dollars for their land and homes toward the construction of new ones in New Revilletown, or "Revilletown Park," about a mile away, on the same road but nearer to the center of the town of Plaquemine. Georgia Gulf built another church, Mt. Zion Baptist Church, in Revilletown Park. Another group of families, including Dickerson's, did not accept the housing swap, received financial settlement, and chose their own homes in the nearby town of Brusly. The municipality of Revilletown no longer exists.

By 1991 all of the families had been relocated, and the buildings of the original Revilletown were razed, but the cemetery remained, surrounded on three sides by Georgia Gulf towers that spewed smoke, toxic vapors, and sometimes flames. Revilletown's residents operated under the belief that Georgia Gulf did not own the graveyard, that they would have perpetual access to their ancestors' graves, and that they could continue to fill the graveyard's empty plots. Twenty years later, however, a dispute with the now greatly enlarged Georgia Gulf would land the question of who owned Revilletown's dead in Judge Best's courtroom.

AFTER THE REVILLETOWN BUYOUT, Georgia Gulf executives took notice of a free Thanksgiving dinner sponsored annually by the Council on Aging and organized by Lester Craig, a Georgia Gulf employee and former Revilletown resident. Georgia Gulf began sponsoring the Thanksgivings, explaining in a 2002 press release:

> Iberville Parish gives us so much, it only makes sense we would want to give something back during the holidays. That's why we sponsor the Council on Aging Thanksgiving Dinner, which provides about 800 local senior citizens with a holiday meal. . . . We see how Iberville Parish residents give freely every year to make holidays better for others. So, when it comes to giving back, Georgia Gulf takes its cue from you. Because that's what neighbors do.
>
> It's what good neighbors do.

Did a few thousand dollars of turkey dinner make Georgia Gulf a good neighbor to the scattered remains of Revilletown? Many residents did not think so, and refused to attend these displays of "neighborliness."

Moreover, amidst the final stages of the town's relocation, Georgia Gulf was among several companies that argued against a proposed EPA standard that would regulate the quantity of dioxins, which cause cancer and immune collapse, in the waste material of vinyl production. In a 1991 study by the National Environmental Law Center and the Center for Public Interest Research, Georgia Gulf remained one of the top dioxin waste culprits. Georgia Gulf's annual report to the Securities and Exchange Commission in 1993 identified fifty-eight hazardous waste disposal sites that were under investigation by the EPA. That same year, the Louisiana Advisory Committee to the U.S. Commission on Civil

Rights issued a report confirming that plants in six towns, including Revilletown, emitted about 2 billion pounds of unregulated toxic waste and known carcinogens during the previous year. The report stated:

> Black communities in the corridor between Baton Rouge and New Orleans are disproportionately impacted by the present state and local government system for permitting and expansion of hazardous waste and chemical facilities. These communities are most often located in rural and unincorporated areas, and residents are of low socioeconomic status with limited political influence.

The report provided encouragement for activists who were eager to demand more than just a buffer zone; they wanted to hold Georgia Gulf accountable for the damage it had done and was doing to the land and people of Iberville Parish. Organizers and activists from Louisiana, including Dickerson, traveled to Washington, D.C., to testify in a congressional hearing on environmental justice. They spoke about the buyouts and sealed settlement documents, the ever-increasing pollution, community health issues, and deadly explosions.

On February 11, 1994, President Bill Clinton issued an "Executive Order on Federal Actions to Address Environmental Justice in Minority Populations and Low-Income Populations." It stipulated:

> Each Federal agency shall analyze the environmental effects, including human health, economic, and social effects, of Federal actions, including effects on minority communities and low-income communities, when such analysis is required by the National Environmental Policy Act of 1969. . . . Mitigation measures outlined or analyzed in an environmental assessment, environmental impact statement, or record of decision, whenever feasible, should address significant and adverse environmental effects of proposed Federal actions on minority communities and low-income communities.

The executive order was deemed in strict adherence to Title VI of the Civil Rights Act of 1964. Shortly after the order was issued, the EPA released a document of actionable steps based on the executive order. Yet the toxicity in Iberville Parish continued to grow unchecked. A 1999 report and study from the Committee on Environmental Justice Institute of Medicine confirmed that Georgia Gulf was maintaining its position as a top dioxin producer.

IN EARLY 2012, when Dickerson and Chiphe went to lay flowers on their families' graves in the Revilletown Cemetery, they were surprised to discover a burial taking place. Neither the family nor the deceased were from Revilletown, and the family told Dickerson and Chiphe that they had paid a woman from an unaffiliated church a fee of $600 for a plot in the cemetery. Dickerson and Chiphe's organization, Mt. Zion Baptist Association, which was formed by the descendants of Revilletown's founders, still held the original 1881 deed to the cemetery and did not charge for burials. The $600 was nowhere to be found in the Mt. Zion Baptist Association's coffers. Shirley Oliver, the local woman to whom these $600 checks were being made out, has refused to comment publicly on the matter. Mt. Zion Baptist Association filed a petition on October 8, 2012, for "injunctive relief against the alienation, encumbrance, or destruction of any burial plots," both to stop the unauthorized collection of money for burial and to prevent the burials of those not descended from Revilletown families.

When they next went to the cemetery, they found the gate padlocked and a sign that read, "Visiting Hours: 9:00am– 5:00pm." Soon Georgia Gulf—now Axiall Corporation—filed its own claim on the cemetery as an intervenor in the suit. Its lawyers claimed that the corporation had exercised full ownership of the cemetery parcel since 1970, when the property was purchased for the original Georgia Gulf plant. Though the company did not have the deed, Axiall argued that it legally owned the cemetery and had granted a separate organization, Mt. Zion Baptist Church #1, access rights in a "good faith agreement." Mt. Zion Baptist Church #1 was represented by Shirley Oliver and unaffiliated with the Mt. Zion Baptist Association.

By the time the case came before Judge Best in 2015, Georgia Gulf had acquired PPG Industries, Inc., in a $2 billion deal and renamed itself Axiall Corporation, and thus the case's name, *Mt. Zion Baptist Association v. Mt. Zion Baptist Church #1 (Intervenor: Axiall Corporation)*. For lead counsel, Dickerson and Chiphe hired attorney Jerome D'Aquila. D'Aquila was a veteran on the Iberville Parish court circuit. He had been practicing law long before Best became a judge, and the two had a long history of antagonism and mudslinging documented in the local papers. D'Aquila's three-piece suit hung from his frame unevenly, suggesting the commanding presence of his earlier days.

By contrast, Axiall Corporation was represented by a team of six sharply dressed men. They set up a projector, screen, and speakers, and delivered a multimedia presentation with maps and photos, including one of the grass in the cemetery. The Axiall team had measured the length of the grass that morning to demonstrate that the Mt. Zion Baptist Association failed to maintain the cemetery grounds. Dickerson nearly leapt out of her seat. D'Aquila cut in, "Mt. Zion Baptist Association has sent someone to mow every week, but when they arrive, the gate is locked, and Axiall takes hours to send someone to open it when you call!"

When it was our turn, the six men leaned back in their chairs, whispering to one another. D'Aquila spoke, clutching sheets of his yellow legal pad in one hand and his cane in the other. "The people buried in this cemetery were enslaved! And these are their heirs!" He waved his papers in the air. "You cannot take their ancestors!"

Judge Best ruled in favor of Mt. Zion Baptist Church #1, Intervenor: Axiall Corporation. Dickerson and Chiphe had no legal claim to the cemetery. They would need to provide Axiall two days notice to gain access to their ancestors' graves. We exited the courtroom more quietly than we had entered.

"Do you like fish, honey?" Chiphe asked. "Come get in the car." We drove to Dickerson's cousin's restaurant nearby. The menu board read:

SPECIAL
FISH
CHICKEN

Dickerson's cousin kissed us all and made us sit down with plastic cups of iced tea. A few minutes later, she emerged with four giant pieces of fried catfish, greens, butter yeast rolls, and Jell-O for dessert. I distinctly remember wiping my greasy hands and guiltily thinking that I felt full when I should have been feeling empty.

AFTER JUDGE BEST'S RULING, Dickerson and Chiphe filed an appeal. On October 31, 2016, Judge Best dismissed it. They planned to appeal again, this time to the Louisiana Court of Appeals, First Circuit, in Baton Rouge.

In the meantime, relations between Axiall and Mt. Zion Baptist Association soured even more. Dickerson and Chiphe continued to call plant security and give two days' notice for permission to spend time with their ancestors. (After the drama over the length of the grass, Axiall did not keeping up with the mowing, leaving Mt. Zion Baptist Association to again care for the property.)

Things became particularly dramatic when one of Dickerson's relatives died. The burial was just about to begin when Axiall called the Iberville Parish Sheriff's office, contending that, because they were still in litigation, no one was permitted on the cemetery property. The grave had already been dug. The family members and the truck bearing their dead were escorted out of the cemetery by the sheriff.

The First Circuit eventually ruled that Axiall did not possess the cemetery. Dickerson and the Mt. Zion Baptist Association appealed again in Iberville Parish Court for access. The appeal was dismissed on the grounds that Dickerson and Chiphe had to be elected representatives by members of the association. Since then, elections were held, and Dickerson was confirmed as a representative. And so the case remains in court, with both groups taking legal action against the other. The Mt. Zion Baptist Association swears to never stop fighting.

SOMEDAY MY FRIENDS and neighbors up the river will die, just as we all will. But they may die earlier and more painfully than many others. As if this fate were not bad enough, it is possible that their children will have to call a 1-800 number to schedule a visit with them in Revilletown Cemetery.

But black struggle in Louisiana lives on in more than one form, and other battles for ownership and commemoration of African and African American cemeteries are gaining momentum elsewhere. In the spring of 2018, just downriver from Revilletown in Donaldsonville, Louisiana, Shell Corporation recognized two cemeteries of enslaved Africans and African Americans on the borders of one of its large petroleum refineries. Local historians, archaeologists, and Shell administrators had been working on the project since 2013.

Also in 2018, the state of Louisiana assembled the Slavery Ancestral Burial Grounds Preservation Commission, which includes Dickerson, dedicated specifically to discussions of how best to preserve and memorialize cemeteries such as those in Donaldsonville. There is great optimism as this movement for commemoration, born of many years of unsung shared struggles around the state, begins to move into the public spotlight.

But even with the beautiful ceremonies and the erection of memorial plaques, petrochemical companies such as Shell and Axiall continue to defy EPA regulations and fund more pipelines through ancient wetlands—all while issuing press releases that reek of paternalism. They do not realize that networks of kin and comrades are stronger than metal and steam and carbon. The spirit of black resistance is irrepressible; in Louisiana, an attack on the dead is an attack on the living. All of us, up and down the river, will answer the call for support. Because that is what good neighbors do.

William J. Barber II is the President of Repairers of the Breach, Co-Chair of the Poor People's Campaign, Bishop with the College of Affirming Bishops and Faith Leaders, Visiting Professor at Union Theological Seminary, and coauthor of *Revive Us Again: Vision and Action in Moral Organizing.*

Thomas Baxter is a high school principal in rural Ohio. He publishes this essay under a pseudonym.

Lesly-Marie Buer is an applied medical anthropologist and author of the upcoming book *Rx Appalachia*. She is the data manager at Choice Health Network in Knoxville, Tennessee.

Elizabeth Catte is a writer and public historian based in the Shenandoah Valley, Virginia. She is the author of *What You Are Getting Wrong About Appalachia* and coeditor of *55 Strong: Inside the West Virginia Teachers' Strike.*

Ash-Lee Woodard Henderson is the first black woman Co-Executive Director of the Highlander Research & Education Center and is an active participant in the Movement for Black Lives and Southern Movement Assembly.

Nancy Isenberg is the T. Harry Williams Professor of History at Louisiana State University and author of *White Trash: The 400-Year Untold Story of Class in America.*

Elaine C. Kamarck is a Senior Fellow at the Brookings Institution, the author of *Primary Politics: Everything You Need to Know About How America Nominates Its Presidential Candidates*, and a member of the Democratic National Committee.

Michael Kazin teaches history at Georgetown University and co-edits *Dissent.* Author of *American Dreamers: How the Left Changed a Nation*, he is currently a member of Princeton's Institute for Advanced Study.

Toussaint Losier is Assistant Professor of Afro-American Studies at the University of Massachusetts-Amherst and coauthor of *Rethinking the American Prison Movement*.

Robin McDowell is Ph.D. Candidate in African and African American Studies at Harvard University. She was a Gerald Gill Fellow at Tufts University Center for the Study of Race and Democracy and is currently a Fellow at the History Design Studio at the Hutchins Center for African American Research.

Bob Moser is the author of *Blue Dixie: Awakening the South's Democratic Majority*. For nearly two decades, he has reported on southern politics for publications such as *Rolling Stone* and the *New Republic*.

Hugh Ryan, author of *When Brooklyn Was Queer*, is a historian and curator based in Brooklyn.

Matt Stoller is policy director at the Open Markets Institute. He has written for the *New York Times*, the *Washington Post*, and the *New Republic*, and is author of the forthcoming *Goliath: How Monopolies Secretly Took Over the World*.

Ruy Teixeira is a Senior Fellow at the Center for American Progress. His most recent book is *The Optimistic Leftist: Why the 21st Century Will Be Better Than You Think*.

Makani Themba is Chief Strategist at Higher Ground Change Strategies. She lives in Jackson, Mississippi, and is an active member of the Democratic Visioning Committee, which coordinates and supports Jackson People's Assemblies.

Jessica Wilkerson is an Assistant Professor of History and Southern Studies at the University of Mississippi. She is the author of *To Live Here, You Have to Fight: How Women Led Appalachian Movements for Social Justice*.